Model to Meaning

Our world is complex. To make sense of it, data analysts routinely fit sophisticated statistical or machine learning models. Interpreting the results produced by such models can be challenging, and researchers often struggle to communicate their findings to colleagues and stakeholders. Model to Meaning is a book designed to bridge that gap. It is a practical guide for anyone who needs to translate model outputs into accurate insights that are accessible to a wide audience. This book

- Presents a simple and powerful conceptual framework to interpret the results from a wide variety of statistical or machine learning models.
- Features in-depth case studies covering topics such as causal inference, experiments, interactions, categorical variables, multilevel regression, weighting, and machine learning.
- Includes extensive practical examples in both R and Python using the *marginaleffects* software.
- Is accompanied by comprehensive online documentation, tutorials, and bonus case studies.

Model to Meaning introduces a simple and powerful conceptual framework to help analysts describe the statistical quantities that can shed light on their research questions, estimate those quantities, and communicate the results clearly and rigorously. Based on this framework, the book proposes a consistent workflow that can be applied to (almost) any statistical or machine learning model. Readers will learn how to transform complex parameter estimates into quantities that are readily interpretable, intuitive, and understandable.

Written for data scientists, researchers, and students, the book addresses newcomers seeking practical skills and experienced analysts who are ready to adopt new tools and rethink entrenched habits. It offers useful ideas, concrete workflows, powerful software, and detailed case studies presented using real-world data and code examples.

Vincent Arel-Bundock is Professor at the Université de Montréal, where he teaches political economy and research methods. His research focuses on interpreting statistical models more rigorously and making them more accessible. Vincent is the creator of the widely used *marginaleffects* software package, available for both R and Python.

Model to Meaning
How to Interpret Statistical Models with R and Python

Vincent Arel-Bundock

CRC Press
Taylor & Francis Group
Boca Raton London New York

CRC Press is an imprint of the
Taylor & Francis Group, an **informa** business
A CHAPMAN & HALL BOOK

First edition published 2026
by CRC Press
2385 NW Executive Center Drive, Suite 320, Boca Raton FL 33431

and by CRC Press
4 Park Square, Milton Park, Abingdon, Oxon, OX14 4RN

CRC Press is an imprint of Taylor & Francis Group, LLC

ISBN: 978-1-032-90890-8 (hbk)
ISBN: 978-1-032-90872-4 (pbk)
ISBN: 978-1-003-56033-3 (ebk)

DOI: 10.1201/9781003560333

Typeset in Latin Modern font
by KnowledgeWorks Global Ltd.

Publisher's note: This book has been prepared from camera-ready copy provided by the authors.

Contents

Author

Vincent Arel-Bundock is Professor at the Université de Montréal, where he teaches political economy and research methods.

1

Who is this book for?

1.1 The big picture

Our world is complex. To make sense of it, data analysts routinely fit sophisticated statistical or machine learning models. Interpreting the results
produced by such models can be challenging, and researchers often struggle to
communicate their findings to colleagues and stakeholders. *Model to Meaning*
is a book designed to bridge that gap. It is a practical guide for anyone who
needs to translate model outputs into accurate insights that are accessible to
a wide audience.

Model to Meaning introduces a conceptual framework to help you describe
the statistical quantities that can shed light on your research questions, use
models to estimate those quantities, and communicate the results clearly and
rigorously. Based on this conceptual framework, the book proposes an analysis
workflow which can be applied in consistent fashion to (almost) any model
you need to fit.

The *Model to Meaning* project was conceived to empower a broad range of
people—including data scientists, researchers, and students—who want to
improve their ability to interpret and communicate the results produced by
statistical or machine learning models. It is a book for the novice who seeks
new practical skills and understanding, but also for the seasoned researcher
who is ready to unlearn some old patterns and embrace new tools that can
improve their productivity and impact.

Part I of the book lays the groundwork by encouraging analysts to clearly
define their goals, and by introducing a simple conceptual framework to guide
model interpretation. The key idea that underpins this framework is that
we can often transform the raw parameter estimates obtained by fitting a
model into quantities that are much easier to interpret. Converting results
to a scale that feels natural to our audience can improve transparency and
communication.

Part II explains how the conceptual framework can be operationalized through
quantities of interest and tests, using concrete examples and real-world
datasets. It describes three broad classes of quantities of interest—predictions,

DOI: 10.1201/9781003560333-1

1

counterfactual comparisons, and slopes—shows how to estimate them, and explains how to design appropriate hypothesis tests to answer our research questions.

Part III of the book presents detailed case studies to demonstrate how a consistent workflow can be applied in model-agnostic fashion to various settings: causal inference; experiments; interactions and polynomials; mixed effects models; weighting; categorical outcomes; machine learning; and more. These case studies do not exhaust the range of contexts where the tools and ideas in this book can play an integral role. The website that accompanies this book includes over 30 free chapters with detailed tutorials and notebooks.[1]

The level of technical sophistication required to follow the presentation is modest. Readers familiar with concepts like logistic regression and p values should feel comfortable with most of the material. Some of the case studies in Part III cover more advanced modeling approaches, and extra reading materials are cited when appropriate.

Throughout, explanations are accompanied by detailed code examples in R, with Python translations collected in Appendix II. Readers who are not yet familiar with basic data manipulation commands in R or Python may want to consult an additional reference, such as *Telling Stories with Data* (Alexander, 2023), *R for Data Science* (Wickham et al., 2023), or *Python for Data Analysis* (McKinney, 2022).

Parts of this text were adapted from an article by Arel-Bundock et al. (2024) published in the *Journal of Statistical Software*.[2] I thank my co-authors Noah Greifer and Andrew Heiss for their contributions to that article, the marginaleffects software documentation, and code. I acknowledge the use of large language models as writing and coding aids, and thank the Université de Montréal for funding some software development.

Writing this book would not have been possible without the help and feedback of many friends, marginaleffects users, readers, and contributors. I warmly thank Arthur Albuquerque, Rohan Alexander, Marco Mendoza Aviña, Etienne Bacher, Tyson Barrett, Gábor Békés, Daniel K Berry, Mattan S Ben-Shachar, Timothy Chisamore, Nicholas J Clark, Mark Clements, Simon P Couch, Sam Crawley, Maël Coursonnais, Marcio Augusto Diniz, Michael Donnelly, Brett Gall, Isabella R Ghement, Nadjim Fréchet, Alexander Fischer, Stefan Hansen, Alex Hayes, Karl Ove Hufthammer, Philippe Joly, Adrien Lamarche, Florence Laflamme, Daniel Lüdecke, Grant McDermott, Artiom Matvei, A Jordan Nafa, Reiko Okamoto, Demetri Pananos, Julia M Rohrer, Resul Umit, Roel Verbelen, Matt Warkentin, Johannes Weytjens, Brenton Wiernik, Stephen Wild, and Aaron Zipp. I also thank Christophe B. De Muri, Maria G and Dante G for the art. Merci à Sari, Mailis et Béa, les plus meilleures du monde.

---

[1] https://marginaleffects.com

[2] Like all articles in the JSS, the text is published under a permissive Creative Commons 3.0 license. https://creativecommons.org/licenses/by/3.0/

1.2 Software

The key idea that underpins this book is that the raw parameter estimates obtained by fitting a model can often be transformed into more interpretable quantities. Presenting results in a way that resonates with the audience enhances clarity, communication, and impact.

Unfortunately, computing intuitive statistical quantities, along with standard errors, can be a tedious and error-prone process. Furthermore, whereas many excellent packages exist to fit models, software often behaves in idiosyncratic ways, producing objects with incompatible structures or inconsistent behavior. This makes it difficult for analysts to maintain a consistent workflow across projects.

To address this challenge, this book introduces a free and open source software package, `marginaleffects`, which provides a single point of entry to interpret results from over 100 classes of models in `R` and `Python`. This package simplifies the interpretation process by offering a consistent and powerful user interface, reducing the need for customized code, and minimizing the risk of errors.

Table 1.1 lists the main functions of the `marginaleffects` package. These functions allow analysts to compute a wide range of quantities, grouped into three categories.

1. `predictions` (Chapter 5): This family of functions computes and plots predictions on different scales, at different levels of aggregation.
2. `comparisons` (Chapter 6): This family of functions computes and plots counterfactual comparisons which can characterize the relationships between two or more variables. This broad class of estimands includes contrasts, differences, risk ratios, odds ratios, lift, etc.
3. `slopes` (Chapter 7): This family of functions computes and plots partial derivatives of the outcome equation, commonly called "marginal effects" in econometrics or "trends" in other disciplines.

The `marginaleffects` package includes many more powerful utilities. For example, the `hypotheses()` function allows analysts to conduct hypothesis or equivalence tests on parameter estimates, or on any of the other quantity produced by the package. This makes it easy to make cross-group comparisons, compare different effect sizes, and more. `datagrid()` is a convenient function to create grids of predictor values; `inferences()` implements alternative inferential strategies like the bootstrap; and `get_draws()` makes it easy to extract draws from posterior distributions in Bayesian analyses.

The functions in `marginaleffects` greatly simplify the analysis of randomized experiments, and can play a key role in analyzing observational data. They are

Table 1.1: Main functions of the `marginaleffects` package.

Goal	Function
Predictions	`predictions()`
	`avg_predictions()`
	`plot_predictions()`
Comparisons	`comparisons()`
	`avg_comparisons()`
	`plot_comparisons()`
Slopes	`slopes()`
	`avg_slopes()`
	`plot_slopes()`
Grids	`datagrid()`
Hypotheses and Equivalence	`hypotheses()`
Bayes, Bootstrap, Simulation	`get_draws()`
	`inferences()`

available in two programming languages, and compatible with over 100 classes of models—more than any comparable package. Supported models include linear, generalized linear (GLM), generalized additive (GAM), mixed-effects, fixed-effects, Bayesian models, and more.

Writing this book was only possible thanks to the work of many developers, including R Core Team (2022), Richardson et al. (2025), Bürkner (2017), Lüdecke, Waggoner and Makowski (2019), Dowle and Srinivasan (2022), Blair et al. (2024), Bergé (2018), Zeileis and Croissant (2010), Kay (2023), Wickham (2016), Brooks et al. (2017), Müller (2020), Xie (2025), Venables and Ripley (2002a), Arel-Bundock (2022), Genz and Bretz (2009), Csárdi and Mühleisen (2024), Venables and Ripley (2002b), Pedersen (2024), Eddelbuettel et al. (2025), Bates and Eddelbuettel (2013), Henry and Wickham (2023), Lang (2017), Allaire et al. (2022), Zeileis et al. (2020), Kuhn and Wickham (2020), Arel-Bundock (2025), and Seabold and Perktold (2010).

1.3   Documentation

The `marginaleffects` package is accompanied by extensive documentation, available both online and in manual pages. In an R session, users can access the manual pages for any function using the standard help syntax.

```
?predictions
?comparisons
?slopes
```

In `Python`, documentation is available through the built-in help system.

```
help(predictions)
help(comparisons
help(slopes)
```

Comprehensive documentation is also hosted on the package website, which includes detailed function references, tutorials, and examples for both R and `Python` users. This website is regularly updated with new features and use cases.

https://marginaleffects.com

1.4   Data

All datasets used in this book can be accessed with the `get_dataset()` function from the `marginaleffects` package. This function can download data from two sources: the `marginaleffects` archive contains datasets specifically curated for this book and the *Rdatasets* archive is a collection of over 2500 datasets which are commonly used in R packages and statistical education.

For example, the code that follows downloads data about the Titanic and displays the first few cells of information.

```
library(marginaleffects)
dat = get_dataset("Titanic", "Stat2Data")
dat[1:5, c("Name", "Survived", "Age")]
```

	Name	Survived	Age
1	Allen, Miss Elisabeth Walton	1	29.00
2	Allison, Miss Helen Loraine	0	2.00
3	Allison, Mr Hudson Joshua Creighton	0	30.00
4	Allison, Mrs Hudson JC (Bessie Waldo Daniels)	0	25.00
5	Allison, Master Hudson Trevor	1	0.92

We can also search through available datasets using plain strings or regular expressions. To find all datasets related to the Titanic, we use the `search` argument.

```
get_dataset(search = "Titanic")
```

Each dataset comes with detailed documentation that you can view in your browser.

```
get_dataset("Titanic", "Stat2Data", docs = TRUE)
```

The datasets used in the *Model to Meaning* book can also be downloaded in CSV and Parquet formats at this URL:

https://marginaleffects.com/data/model_to_meaning.zip

Part I

Interpretation

2

Models and meaning

The best way to start a data analysis project is to set clear goals. This chapter explores four of the main objectives that data analysts pursue when they fit statistical models or deploy machine learning algorithms: model description, data description, causal inference, and out-of-sample prediction.

To achieve these goals, it is crucial to articulate well-defined research questions, and to explicitly specify the statistical quantities—the estimands—that can shed light on those questions. Ideally, estimands should be expressed in the simplest form possible, on a scale that feels intuitive to stakeholders, colleagues, and domain experts.

This chapter concludes by discussing some of the challenges that arise when trying to make sense of complex models. In many cases, the parameters of our models do not directly align with the estimands that actually interest us. Often, we must transform parameter estimates into quantities that directly inform our research questions, and that our audience will readily understand.

2.1 Why fit a model?

The first challenge that all researchers must take on is to transparently state what they hope to achieve with an analysis. Below, I survey four of the goals that an analyst can pursue: model description, data description, causal inference, and out-of-sample prediction.

2.1.1 Model description

The primary aim of model description is to understand how a fitted model behaves in different scenarios. Here, the focus is on the internal workings of the model itself, rather than on making predictions or inferences about the sample or population. The analyst peeks inside the "black-box" to audit, debug, or test how the model reacts to different inputs.

Model description aligns closely with the concepts of interpretability, explainability, and transparency in machine learning. It is an important activity, as it can provide some measure of reassurance that a fitted model works as intended, and is suitable for deployment.

To describe a fitted model's behavior, the analyst might compute model-based predictions, that is, the expected value of the outcome variable for different subgroups of the sample.[1] For example, a real estate analyst may check if their fitted model yields reasonable expectations about home prices in different markets, and go back to the drawing board if those expectations are unrealistic.

In a different context, the analyst may conduct a counterfactual analysis to see how a model's predictions change when we alter the values of some predictors.[2] For instance, a financial analyst may wish to guard against algorithmic discrimination by comparing what their model says about the default risk of borrowers from different ethnic backgrounds, when we hold other predictors constant.

2.1.2   Data description

Data description involves using statistical or machine learning models as tools to describe a sample, or to draw descriptive inference about a population. The objective here is to explore and understand the characteristics of the data, often by summarizing their (joint) distribution. Descriptive and exploratory data analysis can help analysts uncover new patterns, trends, and relationships. It can stimulate the development of theory or raise new research questions (Tukey, 1977).

Data description is arguably more demanding than model description, because it imposes additional assumptions. Indeed, if the sample used to fit a model is not representative of the target population, or if the estimator is biased, descriptive inference may be misleading.

To describe their data, an analyst might use a statistical model to compute the expected value of an outcome for different subgroups of the data.[3] For example, a researcher could offer a summary description of forest coverage in the *Côte Nord* or *Cantons de l'Est* regions of Québec.

2.1.3   Causal inference

In causal inference our goal is to estimate the effect of an intervention (a.k.a., treatment, explanator, independent variable, or predictor) on some outcome (a.k.a., response or dependent variable). This is a fundamental activity in all

---

[1]Chapter 5.

[2]Chapter 6.

[3]Chapter 5

fields where understanding the consequences of a phenomenon helps us make informed decisions or policy recommendations.

Causal inference is one of the most ambitious and challenging tasks in data analysis. It typically requires careful experimental design or statistical models that can adjust for all confounders in observational data. Several theoretical frameworks have been developed to help us understand the assumptions required to make credible claims about causality. Here, I briefly highlight two of the most important.

The structural causal models approach, developed by Judea Pearl and colleagues, encodes causal relationships as a set of equations that describe how variables influence each other. These equations can be represented visually by a directed acyclic graph (DAG), where nodes correspond to variables and arrows to causal effects. Consider Figure 2.1, which shows the causal relationships between a treatment X, an outcome Y, and two confounding variables Z and W. Graphs like this one help us visualize the theorized data generating process and communicate our assumptions transparently. By analyzing a DAG, we can also learn if a research design and statistical model fulfill the conditions required to identify a causal effect in a given context.[4]

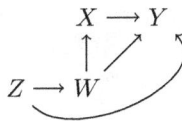

Figure 2.1: Example of a directed acyclic graph.

Another influential theoretical approach to causality is the potential outcomes framework, or Neyman-Rubin Causal Model. In this tradition, we think of causal effects as comparisons between potential outcomes: what would happen to the same individual under different treatment conditions? For each observation in a study, there are two potential outcomes. First, we ask what would happen if unit i received the treatment. Then, we ask what would happen if unit i did *not* receive the treatment. Finally, we define the causal effect as the difference between those two potential outcomes. Unfortunately, we can only observe one potential outcome at a time, because an individual cannot be part of both the treatment and control groups simultaneously. As a result, it is impossible to directly observe the effect of a treatment on any single unit of observation.[5] This inconvenient fact is often dubbed the "fundamental

---

[4]One way to buttress claims of causal identification is to argue that a model satisfies the "backdoor criterion" (Pearl, 2009). A set of control variables $\mathbf{K}$ fulfills this criterion relative to a pair of variables (X, Y) if no node in $\mathbf{K}$ is downstream in the causal path from the treatment X; and if elements of $\mathbf{K}$ "block" every path between X and Y that contains an arrow into X.

[5]Note that repeated measurements do not solve this problem, since a person is not exactly identical at times t and $t + 1$.

problem of causal inference," and it motivates the development of procedures to estimate aggregated or average treatment effects (see Chapters 6 and 8) .

The literature on causality is vast, and a thorough discussion of Structural Causal Models or the Neyman-Rubin Model lies outside the scope of this book.[6] Nevertheless, it is useful to reflect a bit on the language that data analysts use when discussing their empirical results.

When a research design does not satisfy the strict requirements for causal identification, researchers will often revise their texts, replacing strong causal words by weak associational terms. For instance, they may strike out words like "cause", "affect", "influence", or "increase" from observational studies, and replace them by terms such as "link", "association", or "correlation." Paying attention to the words we use to describe statistical results is one way to remind our audience that association does not necessarily imply causation (Hill et al., 2024).

In recent years, though, some have started to push back against the use of causal euphemisms (Grosz et al., 2020; Hernán, 2018). The problem is that researchers and their audiences often care about causation, not mere association, and that pretending otherwise may be counterproductive. Consider the large number of studies published on the "association between daily alcohol intake and risk of all-cause mortality" (Zhao et al., 2023). The research question that motivates this work is fundamentally causal: readers hope to learn how much longer they might live if they stopped drinking, and policy makers wish to know if discouraging alcohol intake would be good for public health. We do not care about mere association; we care about the effects of a cause!

Changing a few adjectives in a paper does not fundamentally change the research question, and it does not absolve us of the responsibility to justify our assumptions. In fact, avoiding causal language may introduce ambiguity about a researcher's goals, and could hinder scientific progress.

What are we to do? One way out of this conundrum is transparency. Researchers can candidly state: "I am interested in the causal effect of X on Y, but my research design and data are limited, so I cannot make strong causal claims." When both the goals of an analysis and the violations of its assumptions are laid out clearly, readers can interpret (and discount) the results appropriately.

Part II of this book presents two broad classes of estimands which can be used to characterize either the "association" between two variables, or the "effect" of one variable on another (Chapters 6 and 7). When describing those quantities, I will not shy away from using causal words like "increase," "change," or "affect." However, analysts should be mindful that when conditions for causal

---

[6]Readers who need to know if their statistical results can be interpreted as causal estimates are encouraged to read one of the many excellent books published on the topic, e.g., Pearl (2009), Angrist and Pischke (2009), Morgan and Winship (2015), Imbens and Rubin (2015), Pearl and Mackenzie (2018), Hernán and Robins (2020), or Ding (2024).

identification are violated, they have a duty to signal those violations explicitly and clearly to their audience.

2.1.4 Out-of-sample prediction

Out-of-sample prediction and forecasting aim to predict future or unseen data points based on a model fit on existing data. This is particularly challenging due to the need to ensure that our model does not overfit the data and generalizes well (James et al., 2021). Furthermore, out-of-sample prediction imposes additional requirements on the stability of the data distribution, and on the absence of changes in exogenous factors between the training sample and the target data. For example, a model trained to predict the probability of loan default may no longer perform well after an economic crisis changes the personal finances of a large portion of the population.

Out-of-sample prediction is not the main focus of this book, but Section 14.4 presents a case study on conformal prediction, a flexible strategy to make predictions and build intervals that will cover a specified proportion of out-of-sample observations.

2.2 What is your estimand?

Once the overarching goal of an analysis is posed, the next step is to rigorously define the target of our inquiry, that is, the specific value that would shed light on our research question. In this book, I will call this target the "quantity of interest" or the "estimand."

An *estimand* is the quantity or parameter that we seek to learn. An *estimator* is the statistical method, algorithm, or mathematical formula that we apply to our data to gain insight into the estimand. An *estimate* is the numerical result that we obtain by applying an estimator to data; it is our best guess of the estimand's true value based on the available information. For example, if we want to know the average height of all adults in a country (estimand), we might use the mean formula (estimator) to calculate that the average height is 170cm in a random sample from the population (estimate).

In statistical training and practice, much emphasis is placed on the various estimators that one can deploy, and on the interpretation of parameter estimates. Estimands do not always get as much attention. This is unfortunate, because a clear definition of the quantity of interest is essential to ensure that the estimator we choose, and the estimates we compute, speak to our research question. As Lundberg et al. (2021) write:

"In every quantitative paper we read, every quantitative talk we attend, and every quantitative article we write, we should all ask one question: what is the estimand? The estimand is the object of inquiry—it is the precise quantity about which we marshal data to draw an inference."

This injunction is sensible, yet many of us conduct statistical tests without explicitly defining the quantity of interest, and without explaining exactly how this quantity answers the research question. This lack of clarity impedes communication between researchers and audiences, and it can lead us all astray.

To see how, imagine a dataset produced by the same data generating process as in Figure 2.1. Further assume that relationships between all variables are linear. In that context, we can estimate the causal effect of X and Y by fitting a linear model.[7]

$$Y = \beta_1 + \beta_2 X + \beta_3 Z + \beta_4 W + \varepsilon \qquad (2.1)$$

At first glance, it may seem that this model allows us to simultaneously and separately estimate the effects of X, Z, and W on Y. Unfortunately, the coefficients in Equation 2.1 do not have such a straightforward causal interpretation.

On the one hand, the model includes enough control variables to eliminate confounding with respect to X.[8] As a result, $\hat{\beta}_2$ can be interpreted causally, as an estimate of the effect of X on Y. On the other hand, Equation 2.1 includes a control variable that is post-treatment with respect to Z: the model adjusts for W, which lies downstream on the causal path from Z and Y. As a result, $\hat{\beta}_3$ should *not* be interpreted as capturing the total causal effect of Z on Y (Keele et al., 2020; Cinelli et al., 2024).[9]

This illustrates what Westreich and Greenland (2013) call the "Table 2 Fallacy": two similar-looking statistical quantities, estimated by a single regression model, can have very different substantive interpretations. More often than not, the coefficients associated with control variables in a regression model are incidental to the analysis. It is a mistake to interpret them individually, one after the other, as if they each captured a causal effect or correlation of interest. To avoid this trap, researchers must define their estimand explicitly, and only highlight the empirical quantities that are actually relevant to this estimand.

---

[7]If the DAG is correct, adjusting for W in the regression model is sufficient to satisfy the backdoor criterion (Hernán and Robins, 2020). Controlling for Z is not strictly necessary to estimate the effect X on Y, but it may improve the precision of our estimates.

[8]In the language of Pearl and Mackenzie (2018) and Morgan and Winship (2015), the backdoors between X and Y are closed.

[9]In the spirit of mediation analysis, $\hat{\beta}_3$ could be interpreted as an estimate of the "direct" effect of Z on Y, rather than its "total" effect. However, this interpretation requires stringent assumptions that are rarely satisfied in practice (Imai et al., 2011; Pearl, 2014). Likewise, $\hat{\beta}_4$ cannot be interpreted as the total effect of W on Y.

If defining one's estimands is so important, why do many researchers neglect to do so? Part of the explanation may simply be that statistical and causal inference are hard problems. It can often be difficult for researchers to follow a consistent thread through theory development, study design, measurement, and statistical modeling, all the way to estimates. To compound this challenge, the terminology used to describe quantities of interest is, frankly, a big mess. In fact, many widely used technical terms have parallel or outright contradictory meanings.

Consider the popular expression "marginal effect." In some disciplines, like economics and political science, a marginal effect is defined as the slope of the outcome with respect to one of the model's predictors. In that context, the word "marginal" evokes the effect of a small change in one of the model's predictors on the outcome.[10] In contrast, researchers from other disciplines who write the same words usually signal that they are "marginalizing," or taking an average of unit-level estimates.[11] In the former case, "marginal effect" refers to a derivative; in the latter, it refers an integral. The same expression has two exactly opposite meanings!

One of the main goals of this book is to empower researchers to overcome terminological ambiguity, and to help them define estimands clearly and easily. To that end, Chapter 3 introduces a powerful new conceptual framework. By answering five simple questions, analysts can rigorously define their quantities of interest, and communicate their results clearly.

2.3 Making sense of parameter estimates

Consider a typical statistical setting where the analyst asks a research question and selects a model to meet domain-specific requirements. Perhaps the model is designed to capture a salient feature of the data-generating process, achieve a satisfactory level of goodness-of-fit, or control for confounders that could introduce bias in a causal estimate. The analyst fits their model using an estimator, and obtains parameter estimates along with standard errors.

Even if a fitted model is relatively simple, the parameter estimates it generates may not map directly onto an estimand that could inform one's research question; the raw parameters may be very difficult to interpret substantively. For example, an analyst who fits a logistic regression model to a binary outcome will typically get coefficient estimates expressed as log odds ratios, that is, as the natural logarithm of a ratio-of-ratios-of-probabilities. Despite the simple

---

[10]Chapter 7
[11]Section 3.3

nature of this model, its parameters are still expressed as complicated functions of probabilities.

But even the most straightforward probabilities are notoriously challenging to grasp intuitively. Indeed, a substantial body of literature in psychology and behavioral economics documents various biases that distort how individuals perceive and make decisions based on them. In their seminal work, Kahneman and Tversky (1979) argue that people tend to "underweight outcomes that are merely probable in comparison with outcomes that are obtained with certainty." Numerous other biases have been identified over the years, including the "base rate fallacy" and various "conservative biases" (Nickerson, 2004).

If understanding simple probabilities is already difficult, how can we expect colleagues and stakeholders to grasp weird concepts like odds ratios? How should we interpret the estimates produced by fancy regression models with splines and interactions? How can we help our audience understand the results produced by mixed-effects Bayesian models, or by complex machine learning algorithms?

The main contention of this book is that, in most cases, analysts should not focus on the raw parameters of their models. Instead, they should transform those parameters into quantities that make more intuitive sense, and that shed light directly onto their research question. Instead of reporting log odds ratios, analysts who fit logistic regressions should transform coefficients into predicted probabilities, and compare predictions made with different values of the predictors.

In Chapter 3, I will argue that this particular transformation, from logit coefficients to predicted probabilities, is but one example of a much more general workflow. This workflow can be applied consistently, in model-agnostic fashion, to interpret the results of over 100 different classes of statistical and machine learning models. By learning one conceptual framework, and one set of tools, analysts can make sense of an extraordinarily large array of modeling contexts.

3

Conceptual framework

> *A parameter is just a resting
> stone on the road to prediction.*
>
> Philip Dawid[*]

This chapter introduces a conceptual framework to aid the interpretation of a wide variety of statistical and machine learning models. This framework is motivated by a simple argument: instead of focusing on the parameters of a fitted model, analysts should convert those parameters into quantities that make more intuitive sense to readers and stakeholders. By applying *post-hoc* transformations, researchers can easily go from model to meaning.

The proposed workflow is both model-agnostic and consistent. Every single analysis starts from the same place: defining the quantity of interest. To do this, we ask three critical questions:

1. *Quantity:* Do we wish to estimate the level of a variable, the association between two (or more) variables, or the effect of a cause?
2. *Predictors:* What predictor values are we interested in?
3. *Aggregation:* Do we care about unit-level or aggregated estimates?

With a clear description of the estimand in hand, we ask two further questions:

4. *Uncertainty:* How do we quantify uncertainty about our estimates?
5. *Test:* Which hypothesis or equivalence tests are relevant?

These five questions give us a structured way to think about the quantities and tests that matter. They lead us to clear definitions of those quantities, and dispel terminological ambiguity. Importantly, the answers to those questions also point to the specific software commands that one needs to execute in order to run appropriate calculations. The rest of the chapter surveys each of the five questions.

---

[*]Dawid is Emeritus Professor of Statistics at Cambridge University. The quote in epigraph was attributed to him by Stephen Senn (Molak, 2024). It may be a paraphrase of Dawid (1985, p.125).

DOI: 10.1201/9781003560333-3

3.1   Quantity

The parameters of a statistical model are often difficult to interpret, and they do not always shed direct light onto the research questions that interest us. In many contexts, it helps to transform parameter estimates into quantities with a more natural and domain-relevant meaning.

For example, the analyst who fits a logistic regression model obtains coefficient estimates expressed as log odds ratios. For most people, this scale is very difficult to reason about. Instead of struggling with complex amalgams of probabilities, analysts should transform estimates into more intuitive quantities, like predicted probabilities or risk differences.

Section 3.1.1 exposes the theoretical underpinnings of *post hoc* transformations: the plug-in principle and the invariance property maximum likelihood. Empirically-minded readers who are less interested in statistical theory may skip that part of the text.

Section 3.1.2 surveys the three classes of quantities of interest at the heart of this book: predictions, counterfactual comparisons, and slopes. These quantities are introduced briefly here, but given chapter-length treatments in Part II.

3.1.1   Theoretical background

There are two primary theoretical justifications for post-estimation transformations: the plug-in principle and the invariance property of maximum likelihood estimators (MLE).

The intuition behind the plug-in principle is simple. To infer some feature of a population, we can study the same feature in a sample, and plug-in our sample estimate in lieu of the population value (Aronow and Miller, 2019).

To formalize this idea a bit, consider a sample $Y_1, \ldots, Y_n$ drawn from a distribution F. For example, F could represent the distribution of ages in a population, and Y_1 could be the observed age of a single individual. Imagine that we care about a statistical functional[1] ψ of this distribution, $\theta = \psi(F)$. In this context, θ could represent the average age of the population, the distribution variance, or the coefficient of a regression model designed to capture the data generating process for F. Since we do not know the age of every individual in the population, we cannot trace the full distribution F and we cannot compute the true $\psi(F)$ directly.

The plug-in principle says that we do not need to know the exact distribution F to learn something about $\psi(F)$. Instead, we can estimate the quantity

---

[1]A statistical functional is a map from distribution F to a real number or vector.

of interest by studying the empirical distribution[2] of a random sample of n individuals, $\hat{F}_n$. More specifically, the plug-in principle states that if $\theta = \psi(F)$ is a statistical functional of the probability distribution F, and if some mild regularity conditions are satisfied,[3] we can estimate θ using the sample analogue $\hat{\theta} = \psi(\hat{F}_n)$.

As Aronow and Miller (2019) note, the implications for statistical practice are profound. When the number of observations increases, the empirical distribution function will tend to approximate the population distribution function. When the number of observations increases, our estimate $\hat{\theta}$ will tend to approach θ.

These ideas empower us to target a broad array of quantities of interest. θ could be as simple as the distribution mean; θ could be a regression coefficient; or θ could be a more interesting quantity, defined as a function of regression coefficients. More concretely, we can estimate the average age of a population by computing the average age of the individuals in our sample, or we can guess the value of a regression parameter in the population by computing the same parameter in-sample.

The plug-in principle thus justifies the workflow proposed in this book. First, we fit a regression model and obtain coefficient estimates that characterize the empirical distribution of the data. Then, we apply a function to coefficient estimates and transform them into more interpretable quantities. Finally, we interpret those quantities as sample analogues to population characteristics.

A different way to motivate *post hoc* transformations is to draw on the invariance property of MLE. This is a fundamental concept in statistical theory, which highlights the efficiency and flexibility of MLE. The invariance property can be stated as follows: if $\hat{\theta}$ is the MLE estimate of θ, then for any function $\psi(\theta)$, the MLE of $\psi(\theta)$ is $\psi(\hat{\theta})$ (Berger and Casella, 2024, 259). In other words, the desirable properties of MLEs—consistency, efficiency, and asymptotic normality—are preserved under transformation.

This property is useful for practice, because it simplifies the estimation of functions of parameters. For example, let's say that we specify a regression model to capture the effect of nutrition on children's heights. We estimate the coefficient θ of this model via maximum likelihood, and obtain an estimate $\hat{\theta}$. Chapter 5 shows how one could apply a function ψ to the coefficient $\hat{\theta}$, in order to compute the predicted (or expected) value of the outcome variable (height) for a given value of the explanatory variable (nutrition). The invariance property says that if the coefficient estimate $\hat{\theta}$ has the desirable properties of a maximum likelihood estimate, then the prediction $\psi(\hat{\theta})$ inherits those qualities.

In sum, it is often useful to apply post-estimation transformations to the parameter estimates obtained by fitting a statistical model, in order to obtain

---

[2]$\hat{F}_n(x) = \frac{1}{n} \sum_{i=1}^{n} I(X_i \leq x), \quad \forall x \in \mathbb{R}$.

[3]See Aronow and Miller (2019, sec. 3.3.1) for an informal discussion of those conditions and Wasserman (2006) for more details.

more meaningful and interpretable quantities. This kind of *post hoc* transformation is well-grounded in statistical theory, via the plug-in principle and the invariance property of MLE.

3.1.2   Predictions, counterfactual comparisons, and slopes

In this book, we will target three broad classes of estimands: predictions, counterfactual comparisons, and slopes. Part II dedicates a full chapter to each of them, with many concrete examples drawing on real-world datasets. The present section foreshadows that in-depth analysis by briefly introducing each quantity.

Predictions

The first class of transformations to consider is the prediction.

> *A prediction is the expected outcome of a fitted model for a given combination of predictor values.*

Consider the general setting where we fit a statistical model by minimizing some loss function $\mathcal{L}$. For a standard linear regression model, $\mathcal{L}$ could represent the sum of squared errors; for a model estimated by maximum likelihood, $\mathcal{L}$ could be the negative log-likelihood. The coefficient estimates that allow us to minimize loss can be written

$$\hat{\beta} = \arg\min_\beta \; \mathcal{L}(Y, X, \mathbf{Z}; \beta), \tag{3.1}$$

where β is a vector of parameters; Y the outcome variable; X a focal predictor which holds particular scientific interest; and $\mathbf{Z}$ a vector of control variables. In Equation 3.1, the arg min operator simply means that the estimator is designed to find the value of β that minimize $\mathcal{L}$.

Once we obtain estimates of the β parameters, we can use them to compute predictions for particular values of the predictors X and $\mathbf{Z}$. To do this, we plug-in the predictors and estimates into an appropriate function Φ.

$$\hat{Y} = \Phi(X = x, \mathbf{Z} = \mathbf{z}; \hat{\beta}), \tag{3.2}$$

The choice of Φ, of course, depends on the model specification. When fitting a model using ordinary least squares, Φ can be a simple linear combination of coefficients and predictors. When Y is a probability, bounded by 0 and 1, Φ can be the logistic or normal distribution functions.[4]

To make things more concrete, imagine that we wish to describe children's heights (Y) as a function of age (X) and caloric intake (Z). The dependent

---

[4]The former would match a logit regression and the latter a probit regression.

variable is numeric and continuous, so we specify a linear regression model with an intercept and two coefficients. After fitting the model, we obtain the following estimates: the intercept is estimated to be $\hat{\beta}_0 = 60$, the coefficient for age is $\hat{\beta}_1 = 6.5$, and the coefficient for caloric intake is $\hat{\beta}_2 = 0.01$.

Now, let's transform these estimates into a prediction. More specifically, let's use our fitted model to compute the expected height of an 11-year-old child who eats 1800 calories per day. To achieve this, we define Φ as a linear combination of coefficients and predictors, and we plug-in the child's personal characteristics and our parameter estimates into Equation 3.2.

$$\begin{aligned}
\hat{Y} &= \Phi(\text{Age} = 11, \text{Calories} = 1800; \hat{\beta}_0 = 60, \hat{\beta}_1 = 6.5, \hat{\beta}_2 = 0.01) \\
&= \hat{\beta}_0 + \hat{\beta}_1 \cdot \text{Age} + \hat{\beta}_2 \cdot \text{Calories} \\
&= 60 + 6.5 \cdot 11 + 0.01 \cdot 1800 \\
&= 149.5 \text{ cm}
\end{aligned}$$

For this specific combination of predictor values, and given the estimated values of the coefficients, our model predicts a height of 149.5 cm. This example shows how easy it is to transform abstract-seeming parameters into a concrete, meaningful, and interpretable quantity: the expected height of an 11-year-old who consumes 1800 calories a day.

Throughout the book, we will perform these operations repeatedly, applying them to more complex models and various Φ functions. At times, we will generate multiple predictions and aggregate them to obtain an average estimate. In other instances, we will compare counterfactual predictions to estimate treatment effects.

Counterfactual comparisons

If parameters are a resting stone on the way to prediction, predictions are a springboard to comparison and causal inference. The second class of transformations to consider is thus the counterfactual comparison.

A counterfactual comparison is a function of two predictions made with different predictor values.

Suppose we want to evaluate the effect of changing the focal predictor X from 0 (control group) to 1 (treatment group) on the predicted outcome $\hat{Y}$, all else equal. To do this, we compute two predictions: one with $X = 0$ and another with $X = 1$, while keeping all control variables at fixed values $\mathbf{Z} = \mathbf{z}$.

$$\begin{aligned}
\hat{Y}_{X=1, \mathbf{Z}=\mathbf{z}} &= \Phi(X = 1, \mathbf{Z} = \mathbf{z}; \hat{\beta}) && \text{Treatment} \\
\hat{Y}_{X=0, \mathbf{Z}=\mathbf{z}} &= \Phi(X = 0, \mathbf{Z} = \mathbf{z}; \hat{\beta}) && \text{Control}
\end{aligned}$$

These two $\hat{Y}$ predictions are called "counterfactual," because they represent the expected outcomes for hypothetical scenarios where the focal predictor X is different from what we factually observe in the data. Here, the key idea is to use a statistical model to extrapolate, to predict what *would* happen to Y if X took on different values.

The next step in the analysis is to choose a function to compare the relative size of the two counterfactual predictions. One natural choice is to simply subtract one from the other. This gives us the difference in predicted outcome between the treatment and control groups, for individuals with identical background characteristics $\mathbf{z}$.

$$\hat{Y}_{X=1,\mathbf{Z}=\mathbf{z}} - \hat{Y}_{X=0,\mathbf{Z}=\mathbf{z}}$$

If $\hat{Y}_{X=1,\mathbf{Z}=\mathbf{z}} > \hat{Y}_{X=0,\mathbf{Z}=\mathbf{z}}$, our model tells us that being part of the treatment group is associated with higher values of Y, for individuals with background characteristics $\mathbf{Z} = \mathbf{z}$.

Chapter 6 shows that this is but one example of a versatile strategy, which can be extended into at least three directions. First, depending on the model type, the predicted outcome $\hat{Y}$ can be expressed on a variety of scales. In linear regression, $\hat{Y}$ is a real number. Using a logistic regression, one could predict and compare predicted probabilities. With Poisson models, we could compare expected outcomes expressed as counts.[5]

Second, the analyst can select different counterfactual values of the focal predictor X. In the example above, X was fixed to 0 or 1. In Chapter 6, we will use different counterfactual values to estimate how the predicted value $\hat{Y}$ reacts when X changes by 1 or more units (e.g., one standard deviation), or from one category to another.

Third, a counterfactual comparison can be expressed as any function of two predictions. It can be a difference $Y_{X=1,\mathbf{Z}=\mathbf{z}} - Y_{X=0,\mathbf{Z}=\mathbf{z}}$, a ratio $\frac{Y_{X=1,\mathbf{Z}=\mathbf{z}}}{Y_{X=0,\mathbf{Z}=\mathbf{z}}}$, the lift $\frac{Y_{X=1,\mathbf{Z}=\mathbf{z}} - Y_{X=0,\mathbf{Z}=\mathbf{z}}}{Y_{X=0,\mathbf{Z}=\mathbf{z}}}$, or even more complex (and unintelligible) functions like odds ratios.

Counterfactual comparisons are a fundamental building block for causal inference. Indeed, when conditions for causal identification are satisfied, they will often be interpreted as measures of the effect of X on Y.[6]

That said, it is important to emphasize that counterfactual comparisons need not have a causal interpretation. In some contexts, it is perfectly reasonable

---

[5] In generalized linear models, we could also make predictions on the link scale, rather than the response scale. The `marginaleffects` package allows both, but it is usually better to use the response scale, because its interpretation is more intuitive.

[6] Section 2.1.3 and Chapter 8

to interpret them purely in associational terms, as measures of the statistical relationship between two variables X and Y, holding other variables constant.

To sum up, counterfactual comparisons are a broad class of estimands designed to measure the association between two variables, or the effect of one variable on another: contrasts, differences, risk ratios, lift, etc.

Slopes

The third class of transformations to consider is the slope.

A slope is the partial derivative of the regression equation with respect to a focal predictor.

The slope is the rate at which a model's predictions change when we modify a focal predictor by a small amount, while keeping other predictors constant. Slopes fall in the same toolbox as counterfactual comparisons; they help us measure the strength of association between two variables, or the effect of one variable on another, *ceteris paribus*.

In some disciplines like economics and political science, slopes are known as "marginal effects," where the word marginal indicates that we are looking at the effect of a very small change in one of the predictors.[7] Chapter 7 offers a lot more intuition about the nature of slopes or partial derivatives, and it works through several concrete examples.

3.2 Predictors

Predictions, counterfactual comparisons, and slopes are *conditional* quantities, which means that their values typically depend on all the predictors in a model. Whenever an analyst reports one of these statistics, they must imperatively disclose where it was evaluated in the predictor space. They must answer questions like:

Is the prediction, comparison, or slope computed for a 20-year-old student in the treatment group or for a 50-year-old literary critic in the control group?

Answers to questions like this one can be expressed in terms of profiles and grids.

A profile is a specific combination of values for a focal predictor X and a vector of control variables $\mathbf{Z}$. It is the set of predictor values for one observed or hypothetical individual. Profiles can be observed, synthetic, or partially synthetic. An *observed* profile is a combination of predictor values that actually

---

[7]Section 2.2

appears in the sample. It represents the characteristics of an observed individual or unit in our dataset. A *synthetic* profile represents a purely hypothetical individual with representative or interesting characteristics, like sample means or modes. A *partially synthetic* profile combines the two approaches. Here, we take an actual observation from our dataset, and modify a focal predictor to take on some meaningful value.

For example, imagine that we draw one observation with these characteristics from a dataset: {*Age*=25, *City*=Montreal, *Group*=Control}. We can use this information and a fitted model to compute a prediction for this observed individual. Alternatively, we could modify the *Group* value to *Treatment* rather than *Control*, and compute a prediction for the modified, partially synthetic, profile. Doing so would allow us to answer a counterfactual *what if* question: What would have happened to a 25-year-old Montrealer, if they had been part of the treatment group instead of the control group.

A grid is a collection of one or more profiles. For example, it could hold the empirical distribution of predictors, a grid with one row for each of the individuals that we have actually observed. Using this grid, we could compute one prediction (fitted value) for each row of our dataset. Alternatively, we could construct a grid of synthetic profiles, that is, a grid with combinations of predictors that we may not have actually observed, but that hold scientific interest. With such a grid, we could make predictions for a specific type of individual (e.g., 13-year-old boy), under different treatment regimes (e.g., placebo or treatment).

Defining the grid of predictors is a crucial step in model interpretation. Unless we specify a grid, we cannot clearly define the quantity of interest—or estimand—that sheds light on a research question (Section 2.2). By defining a grid, the analyst gives an explicit answer to this question: what kind of people or units do estimates apply to?

In the rest of this section, we use the `datagrid()` function from the `marginaleffects` package to construct and discuss a variety of grids. We will build these grids by reference to a simulated dataset with 10 observations on three variables: numeric, binary and categorical.

```
library(marginaleffects)
library(tinytable)

set.seed(48103)
N = 10
dat = data.frame(
  Num = rnorm(N),
  Bin = rbinom(N, size = 1, prob = 0.5),
  Cat = sample(c("A", "B", "C"), size = N, replace = TRUE)
)
```

The code examples below rely on the pipe operator |> to send the output of datagrid() to the tt() function from the tinytable package. This allows us to display nicely formatted tables for each grid.

3.2.1 Empirical grid

The first type of grid to consider is most obvious. The empirical distribution is simply the observed dataset. It is a grid composed of all the actually observed profiles in our sample.

```
dat |> tt()
```

Num	Bin	Cat
−0.65471	0	A
−0.49825	1	A
0.33231	0	B
1.12147	0	C
−0.06395	1	C
1.5376	1	B
−0.11144	1	B
0.24231	1	C
−0.16921	1	B
0.57632	0	C

Computing predictions on the empirical distribution is common practice in data analysis, as it yields one fitted value for each observation in the sample. When estimating counterfactual comparisons or slopes, analysts will often start with the empirical grid, manipulate one focal predictor, and see how predicted outcomes are affected.

Studying an empirical grid makes most sense when the observed sample is representative of the population that the analyst targets for inference. When working with convenience samples with very different characteristics from the population, it may make sense to use one of the grids described below or to apply weights as described in Chapter 12.

3.2.2 Interesting grid

If the analyst cares about units with specific profiles, they can use the datagrid() function to create customized grids of "interesting" predictor values. This is useful when one wants to compute a prediction or slope for an individual with given characteristics, such as a 50-year-old engineer from Belgium.

By default, `datagrid()` fixes all variables to their means or modes, except for those variables that the analyst has explicitly defined.

```
datagrid(Bin = c(0, 1), newdata = dat) |> tt()
```

Num	Cat	Bin	rowid
0.2312	B	0	1
0.2312	B	1	2

Note that the `Num` variable was fixed to its mean value, and the `Cat` variable to its mode, because neither variable was specified explicitly by the user.

`datagrid()` also accepts functions to be applied to the variables in the original dataset. In R, the `range` function returns a vector of two values: the minimum and the maximum of a variable; the `mean` function returns a single value; and `unique` returns a vector of all unique values in a variable. With the following function call, we get a grid with $2 \times 1 \times 3 = 6$ rows.

```
datagrid(Num = range, Bin = mean, Cat = unique, newdata = dat) |> tt()
```

Num	Bin	Cat	rowid
−0.6547	0.6	A	1
−0.6547	0.6	B	2
−0.6547	0.6	C	3
1.5376	0.6	A	4
1.5376	0.6	B	5
1.5376	0.6	C	6

3.2.3   Representative grid

A representative grid is one where predictors are fixed to representative values, such as means, medians, or modes. This kind of grid is useful when the analyst wants to compute predictions, comparisons, or slopes for a typical or average individual.

```
datagrid(grid_type = "mean_or_mode", newdata = dat) |> tt()
```

Num	Bin	Cat	rowid
0.2312	1	B	1

In Part II of the book, we will see how representative grids allow us to compute useful quantities such as "fitted value at the median" or "slope at the mean." In both cases, the quantities of interest are evaluated at a central point in the predictor space.

Studying representative grids may be useful when looking for a measure of central tendency. It can also be beneficial for computational reasons, because calculating a single statistic for a single profile is cheaper than computing one statistic per row for a large dataset. On the downside, it is important to keep in mind that, in some cases, nobody in the population is exactly average on all dimensions. In that context, the interpretation of quantities computed on representative grids is somewhat ambiguous.

3.2.4 Balanced grid

A balanced grid is a grid built from all unique combinations of categorical variables, and where all numeric variables are held at their means.

To create a balanced grid, we fix numeric variables at their means and create rows for each combination of values for the categorical variables (i.e., the cartesian product). In the next table, `Num` is held at its mean, and rows show all combinations of unique `Bin` and `Cat`.

```
datagrid(grid_type = "balanced", newdata = dat) |> tt()
```

Num	Bin	Cat	rowid
0.2312	0	A	1
0.2312	0	B	2
0.2312	0	C	3
0.2312	1	A	4
0.2312	1	B	5
0.2312	1	C	6

Balanced grids are often used to analyze the results of factorial experiments in convenience samples, where the empirical distribution of predictor values is not representative of the target population. In that context, the analyst may wish to report estimates for each combination of treatments, that is, for each cell in a balanced grid. We will see balanced grids in action in Section 5.3, when computing marginal means.[8]

3.2.5   Counterfactual grid

The last type of grid to consider is counterfactual. Here, we duplicate the entire dataset, creating one copy for every combination of values that the analyst supplies.

In the example that follows, we specify that `Bin` must take on values of 0 or 1. Then, we create two copies of the full dataset. In the first copy, `Bin` is fixed to 0; in the second copy, it is set to 1. All other variables are held at their observed values. Since the original data had 10 rows, the counterfactual dataset has 20 rows.

```
g = datagrid(
  Bin = c(0, 1),
  grid_type = "counterfactual",
  newdata = dat
)
nrow(g)
```

```
[1] 20
```

We can inspect the first three rows of each counterfactual version of the data by filtering on the `rowidcf` column, which holds row indices. Notice how each of the original rows has an exact duplicate, which differs only in terms of the counterfactual `Bin` variable.

```
subset(g, rowidcf %in% 1:3) |> tt()
```

rowidcf	Num	Cat	Bin
1	−0.6547	A	0
2	−0.4982	A	0
3	0.3323	B	0
1	−0.6547	A	1
2	−0.4982	A	1
3	0.3323	B	1

This kind of duplication will be essential in chapters 6 and 8, where we explore counterfactual analysis and causal inference.

---

[8]Balanced grids are used by default in the `emmeans` post-estimation software (Lenth, 2024).

3.3 Aggregation

As noted above, predictions, counterfactual comparisons, and slopes are conditional quantities, in the sense that they typically depend on the values of all predictors in a model. When we compute these quantities over a grid of predictor values, we obtain a collection of different point estimates: one for each row of the grid.

If a grid has many rows, the large number of estimates that we generate can be unwieldy, making it more difficult to extract clear and meaningful insights from our models. To simplify the presentation of results, analysts might want to aggregate unit-level estimates into more macro-level summaries. We can think about several aggregation strategies.

- No aggregation: Unit-level estimates.
- Overall average: Average of unit-level estimates.
- Subgroup averages: Average of unit-level estimates within subgroups of the data.
- Weighted averages: Weighted average of unit-level estimates, where weights could be sampling weights or the inverse probability of treatment assignment.

Aggregated estimates are very common in all fields of research. In Chapter 5, we will compute average predictions or marginal means. In Chapter 6, we will compute average counterfactual comparisons. In Chapter 8, we will see that aggregating comparisons over different grids allows us to compute quantities such as the Average Treatment Effect (ATE), the Average Treatment effect on the Treated (ATT), or the Average Treatment effect on the Untreated (ATU). In Chapter 7, we will compute average slopes (a.k.a. average marginal effects).

Aggregated estimates tend to be easier to interpret, and they can often be estimated with greater precision than unit-level quantities. However, they also come with an important downside, since they can mask interesting variation across the sample. For example, imagine that the effect of a treatment is strongly negative for some individuals, but strongly positive for others. In that case, an estimate of the average treatment effect could be close to 0 because positive and negative estimates cancel out. This aggregated quantity would not adequately capture the nature of the data generating process, and it could lead to misleading conclusions.

3.4 Uncertainty

Whenever we report quantities of interest derived from a statistical model, it is essential to provide estimates of our uncertainty. Without standard errors or confidence intervals, readers cannot confidently assess if the reported values or relationships are genuine, or if they could merely be the product of chance.

The `marginaleffects` package offers four primary methods for quantifying uncertainty: the delta method (default), bootstrap, simulation-based inference, and conformal prediction. Chapter 14 is entirely dedicated to uncertainty quantification. It offers intuition, technical details, and hands-on demonstration for all four of the approaches listed above. Here, we survey them briefly.

The delta method is the default and most common way to compute standard errors for the kinds of quantities considered in this book. It is a statistical technique to approximate the variance of a function of random variables. Since all the quantities of interest described in this book can be seen as transformations of model parameters, the delta method can be used to compute standard errors for predictions, counterfactual comparisons, and slopes. The delta method has many advantages: it is fast, flexible, and it can be paired with robust variance estimates to account for heteroskedasticity, autocorrelation, or clustering. The main disadvantages of the delta method are that it relies on a coarse linear approximation, that it assumes a model's parameters are asymptotically normal, and that it requires functions of parameters to be continuously differentiable.

The second strategy for uncertainty quantification is the bootstrap, a resampling-based technique that generates empirical distributions for estimated quantities by repeatedly sampling from the observed data. The bootstrap is an extremely flexible approach, which can be useful when the delta method's assumptions are not met.

The third method, simulation-based inference, involves drawing simulated coefficients from an assumed distribution, and then using those simulated values repeatedly to compute quantities of interest. This strategy is particularly effective when working with complex models or when estimating functions of multiple parameters. It also provides an intuitive way to visualize uncertainty through predictive distributions.

The fourth method, conformal prediction, is a flexible framework for uncertainty quantification that provides valid *prediction*—rather than *confidence*—intervals under minimal assumptions. Unlike traditional approaches that rely heavily on model-specific assumptions or asymptotic approximations, conformal prediction uses split sample strategies to construct intervals that are guaranteed to cover a given share of out-of-sample data points. This method is particularly appealing for its simplicity and adaptability to complex models.

Chapters 5, 6, and 7 show how we can apply these strategies to construct standard errors and confidence intervals around predictions, comparisons, and slopes.

3.5   Test

Once we have computed a quantity of interest and its standard error, we can finally conduct a test to see if our hypothesis or conjecture is correct. This book covers two important testing procedures: null hypothesis and equivalence tests. These procedures are widely used in data analysis, providing powerful tools for making informed decisions based on model outputs.

Null hypothesis tests allow us to determine if there is sufficient evidence to reject a presumed statement about a quantity of interest. For example, an analyst could check if there is enough evidence to *reject* statements such as:

- *Parameter estimates:* The difference between the first and second regression coefficients is equal to one.
- *Predictions:* The predicted number of goals in a game is two.
- *Counterfactual comparison:* The effect of a new medication on blood pressure is null.
- *Slope:* The association between education and income is zero.

Equivalence tests, on the other hand, are used to provide evidence that the difference between an estimate and some reference set is "negligible" or "unimportant." For instance, we could use an equivalence test to *support* statements such as:

- *Parameter estimates*: The first regression coefficient is practically equivalent to the second regression coefficient.
- *Predictions*: The predicted probability of scoring a goal is not meaningfully different from 2.
- *Counterfactual comparisons*: The effect of a new medication on blood pressure is essentially the same as the effect of an existing treatment.
- *Slope*: The association between education and income is close to zero.

In sum, null hypothesis tests allow us to establish a difference, whereas equivalence tests allow us to establish a similarity or practical equivalence.

The `marginaleffects` package allows us to compute null hypothesis and equivalence tests on raw parameter estimates, (non-)linear combinations of those parameters, as well as on all the quantities estimated by the package:

predictions, counterfactual comparisons, and slopes. Chapter 4 is entirely dedicated to hypothesis and equivalence testing, and the rest of Part II shows how these tests can be applied to other quantities of interest.

3.6   Summary

This chapter was written to address two problems:

a) The parameters of a statistical model are often difficult to interpret.
b) Analysts often fail to rigorously define the statistical quantities (estimands) and tests that can shed light on their research questions.

We can solve these problems by transforming parameter estimates into quantities with a straightforward interpretation and a direct link to our research goals. In practice, this requires us to answer five questions:

1. *Quantity:* Do we wish to estimate the level of a variable, the association between two (or more) variables, or the effect of a cause?
2. *Predictors:* What predictor values are we interested in?
3. *Aggregation:* Do we care about unit-level or aggregated estimates?
4. *Uncertainty:* How do we quantify uncertainty about our estimates?
5. *Test:* Which hypothesis or equivalence tests are relevant?

The analysis workflow implied by these five questions is extremely flexible; it allows us to define and compute a wide variety of quantities and tests. To see this, we can refer to visual aids adapted from Heiss (2022).

The *Data* in Figure 3.1 represents a dataset with five rows and two variables: a numeric and a categorical one. Each of the light gray cells in the first column corresponds to a value of the numeric variable. The second column of the *Data* represents a categorical variable, with gray and light gray cells showing distinct categories.

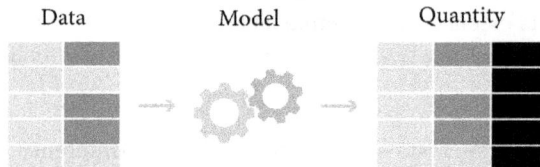

Figure 3.1: Unit-level estimates.

Using these two variables, we fit a statistical model. Then, we transform the parameters of that statistical model to compute quantities of interest. These quantities could be predictions or fitted values.[9] They could be counterfactual comparisons, such as risk differences.[10] Or they could be slopes.[11]

All of these quantities are conditional: their values typically depend on all the predictors in the model. Thus, each row of the original dataset is associated to a different estimate of the quantity of interest. We represent unit-level estimates of the quantity of interest as black cells, on the right-hand side of Figure 3.1.

Often, the analyst does not wish to compute or report one estimate for every row of the original dataset. Instead, they prefer aggregating estimates across the full dataset. For example, instead of reporting one prediction for each row of the dataset, they could first compute individual fitted values, and then marginalize them to obtain an average predicted outcome.

Figure 3.2 illustrates this extended workflow. We follow the same process as above, but add an extra aggregation step at the end to convert unit-level estimates into a one-number summary.

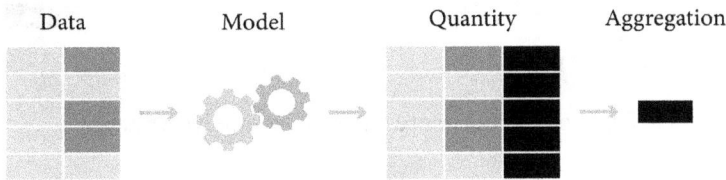

Figure 3.2: Average estimate.

Recall that our *Data* includes a categorical variable, with distinct values illustrated by gray and light gray cells. Instead of aggregating the unit-level estimates across the full dataset, as in Figure 3.2, we could compute average estimates in subgroups defined by the categorical variable. On the right side of Figure 3.3, the first black cell represents an average estimate in the gray subgroup. The second black cell represents the average estimate in the light gray subgroup.

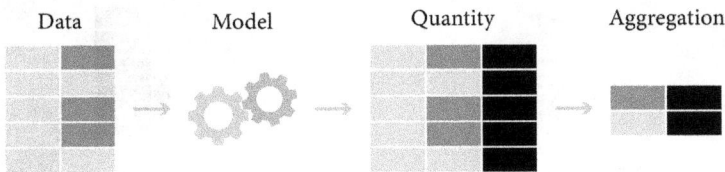

Figure 3.3: Group-average estimates.

---

[9]Chapter 5
[10]Chapter 6
[11]Chapter 7

So far, we have aggregated unit-level estimates computed on the empirical grid, that is, we have taken averages across the set of 10 units of observation. One alternative approach would be to build a smaller grid of predictors, and to compute estimates directly on that smaller grid. The values in that grid could represent units with particularly interesting features. For example, a medical researcher could be interested in computing estimates for a specific patient profile, between the ages of 18 and 35 with a particular biomarker.[12] Alternatively, the analyst could want to compute an estimate for a typical individual, whose characteristics are exactly average or modal on all dimensions.[13]

Figure 3.4 illustrates this approach. Based on the original data, we create a one-row grid which holds the characteristics of a "typical" or "representative" individual. Then, we use the model to compute an estimate—prediction, comparison, or slope—for that synthetic individual.

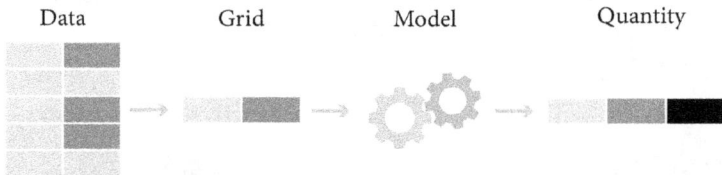

Figure 3.4: Estimates at representative values.

The last approach we consider, illustrated by Figure 3.5, combines both the grid creation and the aggregation steps. First, we duplicate the entire dataset, and we fix one of the variables at different values in each of the datasets. In Section 3.2.5, we described this process as creating a counterfactual grid.

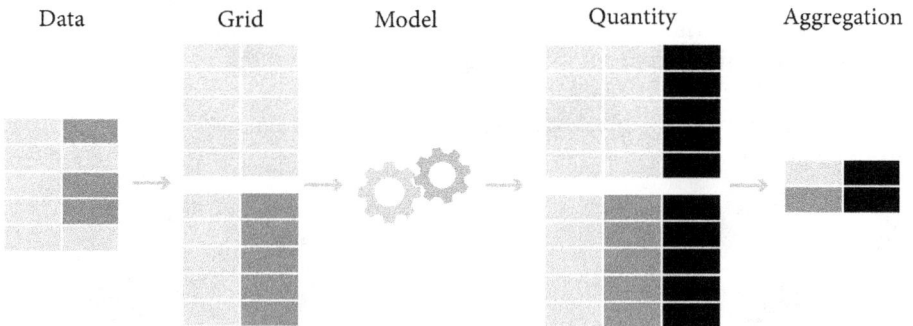

Figure 3.5: Counterfactual average estimates (G-computation).

In the top half of the duplicated grid, the categorical variable is set to a single counterfactual value: light gray. In the bottom half of the duplicated

---

[12]Section 3.2.2
[13]Section 3.2.3

grid, the categorical variable is set to a different counterfactual value: gray. Using this grid and the fitted model, we compute estimates for every row. Finally, we marginalize estimates across subgroups defined by the categorical (counterfactual) variable. In Chapter 8, we will see that Figure 3.5 is a visual representation of the G-computation algorithm.

To sum up, estimands and tests can be defined by focusing on their constituent components: the quantity of interest, predictor grid, aggregation level, uncertainty quantification, and hypothesis testing strategy. Taken together, these components form a flexible framework for data analysis that can be adapted to many scenarios. This conceptual framework is also easy to operationalize via the consistent user interface of the `marginaleffects` package for R and `Python`.

Part II

Quantities and tests

4

Hypothesis and equivalence tests

This chapter introduces two statistical testing procedures: null hypothesis and equivalence tests.[1] As we will see, these complementary technologies are extremely versatile; they can be applied to all the quantities of interest studied in this book.

A null hypothesis test is designed to assess if we can reject the possibility that a population parameter—or function of parameters—takes on a specific value, such as zero. Null hypothesis tests are common in all fields of data analysis. They can help us answer questions like:

1. Does cognitive-behavioral therapy have a non-zero effect on depression?
2. Is the effect of a new drug different from the effect of an existing treatment?
3. Is there a statistically significant difference in test scores between students who attend public or private schools?

Equivalence tests flip the logic around. Instead of establishing a difference, they are designed to make a case for similarity. For example, an equivalence test could show that a drug's estimated effect is "equivalent" to zero, or "not meaningfully different" from the effect of another drug. This approach is useful to answer questions like:

1. Is the effect of a generic drug equivalent to that of the branded version?
2. Is the effect of a marketing campaign on consumption negligible?
3. Are the levels of social trust in two communities similar?

Null hypothesis and equivalence tests are extremely flexible tools. They can be applied to model parameters, or to any of the quantities studied in this book: predictions, counterfactual comparisons, and slopes. When you are done reading Part II of this book, you will not only be able to compute these

---

[1]Wald-style null hypothesis tests are described in most statistical textbooks. Readers who want to learn more about equivalence testing can refer to the book length treatment by Wellek (2010) or to articles by Rainey (2014) and Lakens et al. (2018).

DOI: 10.1201/9781003560333-4

quantities, but also to conduct a wide variety of meaningful statistical tests on them.

But even if we appreciate that hypothesis and equivalence tests are powerful tools, we must also recognize that they are fundamentally limited. In particular, when conducting such tests, the analyst must always keep in mind the distinction between statistical and practical significance. We say that a result is "statistically significant" if it would have been unlikely to occur by pure chance (i.e., sampling variation) in a hypothetical world where the null hypothesis and model hold true. We say that a result has "practical significance" when it has important implications for the real world. Whether a result is practically significant is not dictated by statistical considerations; it depends on the field, the research question, and on theory. Many results are statistically significant without having much practical significance. Often, the magnitude of a treatment effect is distinguishable from zero, but it is too small to be of use to practitioners. In those cases, data analysts will typically report small p values for *both* the null hypothesis and the equivalence tests.

The rest of this chapter explores null hypothesis and equivalence tests, and shows how to execute them with the `marginaleffects` package. The main dataset that we use for illustration comes from Thornton (2008): *The demand for, and impact of, learning HIV status.* For this article, the author conducted a randomized controlled trial to find out if they could encourage people to seek information about their HIV status. They administered HIV tests at home to many people in rural Malawi. Then, they randomly assigned some study participants to receive a small monetary incentive if they were willing to travel to a counseling center, in order to learn the results of their test.

The outcome of interest is a binary variable, `outcome`, equal to 1 if a study participant chose to travel to the center, and 0 otherwise. The treatment is a binary variable, `incentive`, equal to 1 if the participant was part of the treatment group who received an incentive, and 0 if they received no money. The researchers also collected information about people's `distance` from the test center, and a numeric identifier for the `village` in which they live. Finally, our dataset includes `agecat`, a measure of the participants' age in three categories: <18, 18 to 35, and >35.

We use the `get_dataset()` function from the `marginaleffects` package to load the dataset in memory, the `head()` function to extract the first few rows, and the `tt()` function from the `tinytable` package to display results in a good-looking table.

```
library(marginaleffects)
library(tinytable)
dat = get_dataset("thornton")
tt(head(dat))
```

Table 4.1: Six rows of the Thornton (2008) dataset.

village	outcome	distance	amount	incentive	age	hiv2004	agecat
43	1	0.5485229	0	0	14	0	<18
117	0	0.8402644	0	0	14	0	<18
2	0	3.3421636	0	0	15	0	<18
6	0	2.3228946	0	0	15	0	<18
11	0	1.3862627	0	0	15	0	<18
14	0	3.8656266	0	0	15	0	<18

When analyzing these data, Thornton (2008) found that 34% of participants in the control group sought to learn their HIV status. In contrast, a small monetary incentive more than doubled this proportion. Simply put, the intervention proved to be highly successful and cost effective.

Over the next few chapters, we will use the `marginaleffects` package to analyze various aspects of Thornton's data. Here, we ask: Do minors, young adults, and older adults have different propensities to seek information about their HIV status?

To answer this question, let us consider a linear probability model with the binary `outcome` as dependent variable and each level of the `agecat` variable as predictors.

$$\text{Outcome} = \beta_1 \cdot \text{Age}_{<18} + \beta_2 \cdot \text{Age}_{18to35} + \beta_3 \cdot \text{Age}_{>35} + \varepsilon \qquad (4.1)$$

We use the `lm()` function to estimate this model by ordinary least squares, adding `-1` to the formula in order to suppress the usual intercept. We then call `coef()` to extract the vector of coefficient estimates.

```
mod = lm(outcome ~ agecat - 1, data = dat)
coef(mod)

    agecat<18 agecat18 to 35    agecat>35
    0.6718750      0.6787004    0.7277354
```

Because there is no other predictor in the model, and since we intentionally dropped the intercept, the coefficients associated with `agecat` levels measure the average `outcome` in each age category. Indeed, the estimated coefficients printed above are exactly identical to subgroup means calculated using the `aggregate()` function.

```
aggregate(outcome ~ agecat, FUN = mean, data = dat)
```

```
      agecat   outcome
1        <18 0.6718750
2 18 to 35 0.6787004
3        >35 0.7277354
```

At first glance, it looks like the probability that a young adult will seek information about their HIV status is smaller than the probability for older adults: 67.9% for participants between 18 and 35 years old, and 72.8% for those above 35. In the next section, we conduct a formal statistical test of this proposition.

4.1   Null hypothesis

The null hypothesis test is a statistical method used to determine if there is sufficient evidence to reject a presumed statement about a population parameter. The null hypothesis H_0 represents a default or initial claim, usually suggesting no effect or no difference in the parameter of interest. For example, H_0 might state that the mean of a population is equal to a specific value, or that there is no association between two variables.

To conduct a null hypothesis test, we begin by choosing a null hypothesis. The choice of H_0 is a substantive one, not a statistical one. It depends on our domain and research question. After choosing H_0, we must pick a test statistic with known sampling distribution, such as t or Z. This sampling distribution represents the distribution of test statistics that we would observe, across samples, if the null hypothesis were true. We then use observed data to compute the test statistic, and compare it to its assumed distribution under H_0. If the test statistic is extreme, we conclude that if the null were true, we would be very unlikely to observe the data that we did. If the test statistic is extreme, we reject the null hypothesis.

Most statistics textbooks discuss the theory of null hypothesis testing.[2] This section is more practical. It illustrates how to use `marginaleffects` to conduct linear or non-linear tests on model parameters or on functions of those parameters. Throughout, we adopt the standard Wald approach and construct Z test statistics of this form

$$Z = \frac{h(\hat{\theta}) - H_0}{\sqrt{\hat{V}[h(\hat{\theta})]}},$$ (4.2)

---

[2]See for example Cameron and Trivedi (2005, Section 7.2), Aronow and Miller (2019, Section 3.4), Hansen (2022a), Hansen (2022b), and Wasserman (2004).

where $\hat{\theta}$ is a vector of parameter estimates; and $h(\hat{\theta})$ is a function of those estimates, a quantity of interest such as a prediction, counterfactual comparison, or slope; H_0 is the null hypothesis; and $\hat{V}[h(\hat{\theta})]$ is the estimated variance of the quantity of interest.[3]

When $|Z|$ is large, we can reject the null hypothesis that $h(\theta) = H_0$. The intuition is straightforward. First, the numerator of Equation 4.2 measures the distance between the estimated quantity of interest and the null hypothesis. When that distance is large, the observed data is far from the data that would be generated if the null were true. This makes H_0 seem less plausible. Second, the denominator quantifies the uncertainty in our estimate. When that uncertainty is small, our estimate is precise, which puts us in a better position to discriminate against the null hypothesis. In sum, when the numerator is large and/or the denominator is small, the absolute value of Z is large, and we can reject the null hypothesis.

Recall that when we estimated the model in Equation 4.1, we obtained these results:

`summary(mod)`

	Estimate	Std. Error	t value	Pr(>|t|)
agecat<18	0.67188	0.02564	26.20	<0.001
agecat18 to 35	0.67870	0.01233	55.06	<0.001
agecat>35	0.72774	0.01336	54.48	<0.001

By default, the summary functions in R and **Python** report null hypothesis tests against a very specific null hypothesis: that a coefficient is equal to zero. Here, the first coefficient is 0.67188 and the standard errors 0.02564. The test statistic is designed to check if we can reject the null hypothesis that the coefficient for `agecat<18` is equal to zero ($H_0 : \beta_1 = 0$):[4]

$$Z = \frac{\hat{\beta}_1 - H_0}{\sqrt{\hat{V}[\beta_1]}} = \frac{0.67188 - 0}{0.02564} = 26.20 \qquad (4.3)$$

Equation 4.3 shows how to compute the test statistic reported by default by R. Does this test make sense from a substantive perspective? Is it interesting? Do we really need a formal test to reject the null hypothesis that 0% of people below the age 18 are willing to retrieve their HIV test result from the clinic? If the answer to any of those questions is "no," we can easily construct alternative test statistics with the **marginaleffects** package.

---

[3]As described in Chapter 14, the default strategy for null hypothesis tests in **marginaleffects** is to compute standard errors using the delta method. That chapter also explains how to use the bootstrap or simulations instead.

[4]By default, R reports t, which is equivalent to Z in large samples.

4.1.1   Choice of null hypothesis

In our running example, a null hypothesis of zero hardly makes sense. Instead, we should specify a different value of H_0, to compare our results against a more meaningful benchmark. For example, we could ask: Can we reject the null hypothesis that the probability of retrieving one's HIV test result is different from a coin flip?

To answer this question, we use the `hypotheses()` function and its `hypothesis` argument.

```
hypotheses(mod, hypothesis = 0.5)
```

Term	Estimate	Std. Error	z	Pr(>|z|)	2.5%	97.5%
agecat<18	0.672	0.0256	6.7	<0.001	0.622	0.722
agecat18 to 35	0.679	0.0123	14.5	<0.001	0.655	0.703
agecat>35	0.728	0.0134	17.0	<0.001	0.702	0.754

The results show that all three Z statistics are large in absolute terms. Therefore, we can reject the null hypotheses that these coefficients are equal to 0.5. If the true chances of seeking information about HIV status were 50/50, we would be very unlikely to observe data like these.

We would draw the same conclusion by computing Wald-style p values manually, measuring the area under the tails of the test statistic's distribution. In R, the `pnorm(x)` function measures the area under the normal distribution to the left of `x`. The two-tailed p value associated to the first coefficient can thus be computed as

```
# First coefficient
b = coef(mod)[1]

# The standard error is the square root of the diagonal element of the
# variance-covariance matrix
se = sqrt(diag(vcov(mod)))[1]

# The Z statistic for Wald test with null hypothesis of b = 0.5
z = (b - .5) / se

# The p-value is the area under the curve, in the tails of
# the normal distribution beyond |Z|
pnorm(-abs(z)) * 2
```

```
  agecat<18
2.043492e-11
```

p is extremely small, which means that we can reject the null hypothesis of $H_0 : \beta_1 = 0.5$.

4.1.2 Linear and non-linear hypothesis tests

In many contexts, analysts are not solely interested in testing against a simple numeric null hypothesis like 0 or 0.5. Instead, they may wish to compare different quantities to one another. For instance, we can ask if the coefficient associated to the first age category is equal to the coefficient associated to the third age category, $H_0 : \beta_1 = \beta_3$.

To conduct this test, all we need to do is supply an equation-style string to the `hypothesis` argument. The terms of this equation start with b, followed by the position (or index) of the estimate. If we are interested in comparing the first and third coefficients, the equation must include b1 and b3.

```
hypotheses(mod, hypothesis = "b3 - b1 = 0")
```

Hypothesis	Estimate	Std. Error	z	Pr(>|z|)	2.5 %	97.5 %
b3-b1=0	0.0559	0.0289	1.93	0.0534	−0.000808	0.113

This is equivalent to computing the difference between the third and first estimated coefficients.

$$0.7277354 - 0.671875 = 0.0558604$$

Can we reject the hypothesis that the probability of seeking one's HIV result is the same in the <18 and >35 groups? That depends on the threshold of statistical significance that one is willing to accept. The p value shown in the table above is very close to 0.05, a conventional threshold of statistical significance. Whether it makes sense to use that threshold in any given application depends on our tolerance to false positives. If mistakenly rejecting the null has costly consequences, we should pick a more stringent threshold of statistical significance. Otherwise, it may be fine to reject the null even if the p value is not extremely small.

In the test above, we checked if the difference between the two coefficients is equal to 0. Rather than a difference, we could also test against the null hypothesis that the ratio of β_3 to β_1 is equal to 1. If this ratio is greater than one, we know that the probability of seeking one's HIV result is higher in the >35 group than in the <18 group. If the ratio is less than one, we know that the probability of seeking one's HIV result is lower in the >35 group than in the <18 group.

```
hypotheses(mod, hypothesis = "b3 / b1 = 1")
```

Hypothesis	Estimate	Std. Error	z	Pr(>|z|)	2.5 %	97.5 %
b3/b1=1	0.0831	0.0459	1.81	0.0699	-0.00676	0.173

Once again, the results suggest that the estimated probability is higher in the older group, but the *p* value does not quite cross conventional threshold of statistical significance of 0.05. Therefore, a conservative analyst would not reject the null hypothesis that these two probabilities are the same.

The equations supported by the **hypothesis** argument are not limited to simple tests of equality, differences, or ratios. Indeed, the user can write equations with more than two estimates or with various (potentially non-linear) transformations.

```
hypotheses(mod, hypothesis = "b2^2 * exp(b1) = 0")
hypotheses(mod, hypothesis = "b1 - (b2 * b3) = 2")
```

marginaleffects also offers a formula-based interface which acts as a shortcut to some of the more common hypothesis tests. For example, if we want to compute the difference between every coefficient and the "reference" quantity (i.e., the first estimate), we supply a formula with the word "reference" on the right side of the tilde symbol (~) and the word "difference" on the left side.

```
hypotheses(mod, hypothesis = difference ~ reference)
```

Hypothesis	Estimate	Std. Error	z	Pr(>|z|)	2.5 %	97.5 %
(agecat18 to 35) − (agecat<18)	0.00683	0.0285	0.24	0.8104	-0.048936	0.0626
(agecat>35) − (agecat<18)	0.05586	0.0289	1.93	0.0534	-0.000808	0.1125

Now, let's say we want to compare each coefficient to the one that immediately precedes it: the young adults to the minors, and the older adults to the young adults. Further suppose we want to compute ratio of coefficients, instead of differences. We can achieve this by setting **ratio** on the left-hand side, and **sequential** on the right-hand side of the formula.

```
hypotheses(mod, hypothesis = ratio ~ sequential)
```

Hypothesis	Estimate	Std. Error	z	Pr(>|z|)	2.5 %	97.5 %
(agecat18 to 35) / (agecat<18)	1.01	0.0427	0.238	0.81193	0.926	1.09
(agecat>35) / (agecat18 to 35)	1.07	0.0277	2.609	0.00907	1.018	1.13

4.1.3 Multiple comparisons and joint hypothesis tests

The goal of null hypothesis testing is to assess if observed data provide enough evidence to reject a null hypothesis. When conducting a single hypothesis test, the probability of Type I error—falsely rejecting the null hypothesis when it is true—is controlled at a predefined significance level, usually 5%. However, when multiple hypothesis tests are performed, the likelihood of at least one Type I error increases with the number of tests. This phenomenon is known as the multiple comparisons problem.

Statisticians have proposed many procedures to adjust hypothesis tests for multiple comparisons, including the Bonferroni, Holm, and Westfall corrections. The `hypotheses()` function in the `marginaleffects` package can apply many such strategies, and report corrected p values as well as family-wise confidence intervals. All we need to do is use the `multcomp` argument.

```
hypotheses(mod, multcomp = "holm")
```

Term	Estimate	Std. Error	z	Pr(>|z|)	2.5 %	97.5 %
agecat<18	0.672	0.0256	26.2	<0.001	0.611	0.733
agecat18 to 35	0.679	0.0123	55.1	<0.001	0.649	0.708
agecat>35	0.728	0.0134	54.5	<0.001	0.696	0.760

The `hypotheses()` function also supports joint hypothesis tests, via the `joint` and `joint_test` arguments. This allows users to test against the null hypothesis that several quantities of interest are jointly/simultaneously equal to zero. The marginaleffects.com website includes documentation and examples on how to conduct such tests.

4.2 Equivalence

In many contexts, analysts are less interested in rejecting a null hypothesis, and more interested in testing whether an estimate is "equivalent" to some benchmark or interval. For example, medical researchers may wish to determine if the effect of a new drug is similar to that of existing treatments, or if it can be considered "negligible" in terms of "clinical significance." To answer such questions, we can use an equivalence test like the two one-sided test or TOST (Wellek, 2010; Rainey, 2014; Lakens et al., 2018).

An equivalence test is a statistical method used to determine if an estimate is "practically equivalent" to a benchmark, within a specified margin of equivalence.

Whereas traditional significance tests attempt to reject a specific (point) null hypothesis, an equivalence test attempts to reject the null hypothesis that the estimand lies outside an interval of practical equivalence. If we can reject that null hypothesis, we conclude that the quantity of interest is likely to be small or close to the benchmark.

To see how this may work in practice, imagine that taking a well-established course of medication reduces the probability of suffering from a cardiovascular arrest by 9 percentage points. A pharmaceutical company introduces a new medication which, they claim, further reduces the chances of an adverse event. The Québec provincial government must decide if they will refund this more expensive drug. To inform decision-making, government defines an interval of equivalence: if the estimated effect of the new treatment is between 8 and 10 percentage points, both drugs shall be considered "equivalent." The definition of this equivalence interval is not a statistical problem. It is a substantive question that depends on the field, research question, costs, theory, etc.

Figure 4.1 illustrates this situation. The horizontal line represents possible values of the parameter of interest. If the effect of the new drug (θ) falls between 8 and 10 percentage points, it is considered equivalent to the effect of the old drug. In this context, the alternative hypothesis is $H_1 : \theta \in [8, 10]$. The null hypothesis is $H_0 : \theta < 8 \vee \theta > 10$.

We conclude for equivalence when we can reject the null hypothesis that the quantity of interest is far from the benchmark, that is, when we reject the H_0 hypothesis that θ falls in the white areas of Figure 4.1. In other words, if the equivalence test is conclusive, we know that it would be surprising if the effect of the new drug were much different from the effect of the old drug.

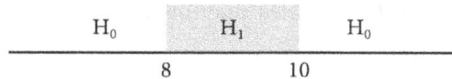

Figure 4.1: In an equivalence test, the null hypothesis H_0 that we attempt to reject is that the quantity of interest lies outside an interval defined by the analyst.

To conduct a TOST of equivalence, we proceed in six steps.

1. *Quantity of interest*: Define and estimate a quantity of interest θ, which can be a coefficient, function of coefficients, prediction, counterfactual comparison, slope, etc.
2. *Significance threshold*: Choose a statistical significance threshold α below which we will reject the null hypothesis.[5]

---

[5] Conventional thresholds include 0.05, 0.01, and 0.001, but these values are arbitrary. The choice should be made based on one's tolerance for false positives in the specific context of the study.

3. *Interval:* Use subject matter knowledge to define an interval of equivalence $[a, b]$. If the quantity of interest θ falls between a and b, it is considered clinically irrelevant or practically equivalent to a benchmark.

4. *Non-inferiority:* Compute the p value associated with a one-tailed null hypothesis test to determine if we can reject the null hypothesis that $\theta < a$.

5. *Non-superiority:* Compute the p value associated with a one-tailed null hypothesis test to determine if we can reject the null hypothesis that $\theta > b$.

6. *Equivalence:* Check if the maximum of the non-inferiority and non-superiority p values is lower than the chosen threshold of statistical significance.

To illustrate, let's revisit the model we fitted above and compare the probability that people in the 18-to-35 and > 35 age brackets will travel to learn their HIV status.

```
coef(mod)
```

agecat<18	agecat18 to 35	agecat>35
0.6718750	0.6787004	0.7277354

```
hypotheses(mod, hypothesis = "b3 - b2 = 0")
```

Hypothesis	Estimate	Std. Error	z	Pr(>|z|)	2.5 %	97.5 %
b3-b2=0	0.049	0.0182	2.7	0.00698	0.0134	0.0847

The results above show that the estimated difference in coefficients for the two groups is equal to 0.0490, and that this difference is statistically significant (i.e., likely different from zero). This difference may be statistically significant, but is it meaningful, clinically relevant, or practically important?

The first step to answer this question is to define exactly what we mean by "meaningful" or "important." Specifically, we must define an interval of equivalence, in which estimates are considered unimportant. There is no purely statistical criterion to construct this interval; the decision depends entirely on domain expertise and subject matter knowledge.

In our running example, the researcher could decide that if the difference in $Pr(\text{outcome} = 1)$ between the young and older adults is between -5 and 5 percentage points, we can ignore it. If the difference falls in the $[-0.05, 0.05]$ interval, it is practically equivalent to zero.

To conduct a TOST on this equivalence range, we simply add the `equivalence` argument to the previous call.

```
hypotheses(mod,
  hypothesis = "b3 - b2 = 0",
  equivalence = c(-0.05, 0.05))
```

Hypothesis	Estimate	Std. Error	p (NonInf)	p (NonSup)	p (Equiv)
b3-b2=0	0.049	0.0182	<0.001	0.479	0.479

These results allow us to reach three main conclusions:

1. *Non-inferiority*: The p value associated to this test is very small ($p < 0.001$). We can reject the null hypothesis that the difference between coefficients is lower than -0.05.
2. *Non-superiority*: The p value associated to this test is large (0.479). We cannot reject the null hypothesis that the difference between coefficients is larger than 0.05.
3. *Equivalence*: The p value associated to the TOST of equivalence corresponds to the maximum of the non-superiority and non-superiority values: 0.479. Again, we cannot reject the null hypothesis that the two coefficients are meaningfully different from one another. We cannot reject the null hypothesis that $\beta_3 - \beta_2 < -0.05 \vee \beta_3 - \beta_2 > 0.05$.

In this example, we applied a TOST to a difference between two coefficients, but the same procedure can be applied to other quantities of interest, such as predictions, counterfactual comparisons, and slopes. The bounds of the equivalence interval can also be set wherever the analyst prefers. Often, the equivalence interval will be centered around zero, but it can be set elsewhere.

4.3   Summary

This chapter introduced two classes of statistical testing procedures: null hypothesis and equivalence tests.

A null hypothesis test allows us to determine if there is enough evidence to reject the hypothesis that a parameter (or function of parameters) is *equal* to a given value.

Examples of statements that could be rejected by a null hypothesis test include:

- The predicted wages of college and high school graduates are equal.
- The effect of a new drug on a health outcome is zero.
- A marketing campaign has the same effect on sales in rural or urban areas.

When a null hypothesis test indicates that we can reject statements like these (small p value), *we establish a difference.*

An equivalence test allows us to determine if there is enough evidence to reject the hypothesis that a parameter (or function of parameters) is *meaningfully different* from a benchmark.

Examples of statements that could be rejected by an equivalence test include:

- The difference in wages between college and high school graduates is considerable.
- The effect of a new drug on a health outcome is meaningfully different from the effect of an existing treatment.
- The effect of a marketing campaign on consumption is much larger than zero.

When an equivalence test indicates that we can reject statements like these (small p value), *we establish a similarity.*

All the main `marginaleffects` functions include `hypothesis` and `equivalence` arguments. These arguments make it easy to conduct null hypothesis and equivalence tests on any of the quantities estimated by the package—predictions, counterfactual comparisons, and slopes—as well as on arbitrary functions of those quantities.

5

Predictions

The parameter estimates obtained by fitting a statistical model are rarely the main object of interest in a data analysis. Instead of focusing on those raw estimates, a good starting point is often to compute model-based predictions for different combinations of predictor values. This allows an analyst to report results on a scale that makes intuitive sense to their readers, colleagues, and stakeholders.

What is a model-based prediction? In this book, we consider that

A prediction is the outcome expected by a fitted model for a given combination of predictor values.

This definition is in line with the familiar concept of "fitted value," but it differs from a "forecast" or "out-of-sample prediction" (Hyndman and Athanasopoulos, 2018). For our purposes, the word "prediction" need not imply that we hope to forecast the future, or that we are trying to extrapolate to unseen data. It simply denotes the best guess of a fitted model for specific predictor values.

Model-based predictions are often the main quantity of interest in a data analysis. They allow us to answer a wide variety of questions, such as:

- What is the expected probability that a 50-year-old smoker develops heart disease, adjusting for diet, exercise, and family history?
- What is the expected probability that a football team wins a game, considering the team's recent performance, injuries, and opponent strength?
- What is the expected turnout in municipal elections, accounting for national trends and local demographic characteristics?
- What is the expected price of a three-bedroom house in a suburban area, controlling for floor area and market conditions?

All of these descriptive questions can be answered using model-based predictions. Predictions are an intrinsically interesting quantity. In Chapters 6 and 7, we will see that they are also a fundamental building block to analyze the effects of interventions, or the association between variables.

The current chapter illustrates how to compute and report predictions for models estimated with the Thornton (2008) data. We proceed in order, through

DOI: 10.1201/9781003560333-5

each component of the conceptual framework laid out in Chapter 3: (1) quantity, (2) predictors, (3) aggregation, (4) uncertainty, and (5) tests. The chapter concludes by showing different ways to visualize predictions.

5.1 Quantity

To begin, it is useful to see how predictions are built in one particular case. Consider a logistic regression model estimated using the Thornton (2008) HIV dataset.

$$Pr(O = 1) = g\left(\beta_1 + \beta_2 \cdot I + \beta_3 \cdot A_{18-35} + \beta_4 \cdot A_{>35}\right), \qquad (5.1)$$

where O is a binary variable equal to 1 if the study participant traveled to the test center to learn the outcome of their HIV test; I is a binary variable equal to 1 if the participant received a monetary incentive; and the other two predictors are indicators for the age category to which a participant belongs, with omitted category $A_{<18}$.

The letter g represents the standard logistic function $g(x) = \frac{1}{1+e^{-x}}$. Applying this function ensures that the linear component inside the parentheses of Equation 5.1 gets rescaled to the $[0, 1]$ interval in order to respect the natural (probability) scale of the binary outcome.

Now, we load the `marginaleffects` package, read the Thornton data, and estimate a logistic regression model using the `glm()` function.

```
library(marginaleffects)
dat = get_dataset("thornton")
mod = glm(outcome ~ incentive + agecat, data = dat, family = binomial)
```

The estimated coefficients are:

```
b = coef(mod)
b
```

```
  (Intercept)       incentive agecat18 to 35        agecat>35
  -0.78232923      1.99229719     0.04368393       0.24780479
```

For clarity of presentation, we substitute these estimates into the model equation.

$$Pr(O = 1) = g\left(-0.782 + 1.992 \cdot I + 0.044 \cdot A_{18-35} + 0.248 \cdot A_{>35}\right) \quad (5.2)$$

To make a prediction for a particular individual, we simply plug-in the characteristics of that person into Equation 5.2. For example, the predicted probability that `outcome` equals 1 for an 18 to 35-year-old in the treatment group is

$$Pr(O = 1) = g\left(-0.782 + 1.992 \cdot 1 + 0.044 \cdot 1 + 0.248 \cdot 0\right)$$

The predicted probability that *Outcome* equals 1 for someone above 35 years-old in the control group is

$$Pr(O = 1) = g\left(-0.782 + 1.992 \cdot 0 + 0.044 \cdot 0 + 0.248 \cdot 1\right)$$

These expressions can be evaluated in two steps. First, we compute the linear—or link scale—component by evaluating the expression inside the parentheses.

```
linpred_treatment_younger = b[1] + b[2] * 1 + b[3] * 1 + b[4] * 0
linpred_treatment_younger
```

```
[1] 1.253652
```

```
linpred_control_older = b[1] + b[2] * 0 + b[3] * 0 + b[4] * 1
linpred_control_older
```

```
[1] -0.5345244
```

Link scale predictions from a logit model are expressed on the log odds scale. In this example, they include a negative value and a value greater than one. To many, this will feel incongruous, because the outcome is a binary variable, with a probability bounded by 0 and 1.

The second step of the computation is thus to apply the logistic function g, to ensure that predictions respect the natural scale of the data.[1]

```
g = \(x) 1 / (1 + exp(-x))
g(linpred_treatment_younger)
```

```
[1] 0.7779314
```

```
g(linpred_control_older)
```

```
[1] 0.3694623
```

---

[1] Instead of defining our own `g()` function, we could use the built-in `plogis()` function in R, or `scipy.stats.logistic.cdf()` in Python.

Our model expects that the probability of seeking information about one's HIV status is 78% for a young adult who receives a monetary incentive, and 37% for an older participant who does not receive an incentive.

Computing predictions manually is useful for pedagogical purposes, but it is labor-intensive and error-prone. The commands above are also limiting, because they only apply to one very specific model.

Instead of manual computation, we can use the `predictions()` function from the `marginaleffects` package. This function can be applied in consistent fashion across more than 100 different classes of statistical models.

First, we build a data frame of predictor values—a grid—where each row represents a different individual.

```
grid = data.frame(agecat = c("18 to 35", ">35"), incentive = c(1, 0))
grid
```

```
     agecat incentive
1 18 to 35         1
2      >35         0
```

Then, we call the `predictions()` function, using the `newdata` argument to specify the predictor values, and the `type` argument to set the scale (link or response):

```
predictions(mod, newdata = grid, type = "link")
```

Estimate	Std. Error	z	Pr(>|z|)	2.5 %	97.5 %
1.254	0.0691	18.15	<0.001	1.118	1.389
−0.535	0.1013	−5.28	<0.001	−0.733	−0.336

These results are exactly identical to the link scale predictions that we computed manually above. This is reassuring but, in a logit model, link scale predictions are hard to interpret.

To communicate our results clearly, it is usually best to make predictions on the same scale as the outcome variable. Doing this makes our estimates easier to interpret, since they can be compared directly to observed values of the outcome variable. For this reason, the default behavior of `predictions()` is to return predictions on the response scale.

```
predictions(mod, newdata = grid)
```

Estimate	Pr(>|z|)	2.5 %	97.5 %
0.778	<0.001	0.754	0.800
0.369	<0.001	0.325	0.417

In the rest of this chapter, we show that the `marginaleffects` package makes it easy to compute various types of predictions, aggregate, and conduct statistical tests on them.

5.2   Predictors

Predictions are *conditional* quantities, in the sense that they typically depend on the values of all the predictor variables in a model. To compute a prediction, the analyst must fix all the variables on the right-hand side of the model equation, that is, they must choose a grid.

The choice of grid depends on the researcher's goals. The profiles it holds could correspond to actual observations in the original data, or they could represent unseen, hypothetical, or representative units. To illustrate, let's consider a slight modification of the model estimated in Section 5.1. In addition to the `incentive` and `agecat` predictors, we now include a numeric predictor to account for the `distance` between a study participant's home and the test center where they can learn their HIV status.

```
mod = glm(outcome ~ incentive + agecat + distance,
    data = dat, family = binomial)
```

With this model, we can make predictions on various grids: empirical, interesting, representative, balanced, or counterfactual.[2]

5.2.1   Empirical grid

By default, the `predictions()` function uses the full original dataset as a grid, that is, it uses the empirical distribution of predictors.[3] This means that `predictions()` will compute fitted values for each one of the rows in the dataset used to fit the model.

```
p = predictions(mod)
p
```

---

[2]Section 3.2
[3]Section 3.2.1

Estimate	Pr(>|z|)	2.5 %	97.5 %
0.365	<0.001	0.297	0.439
0.354	<0.001	0.288	0.426
0.265	<0.001	0.209	0.330
0.300	<0.001	0.241	0.365
0.334	<0.001	0.271	0.402
2815 rows omitted			
0.833	<0.001	0.809	0.855
0.840	<0.001	0.815	0.862
0.833	<0.001	0.809	0.855
0.827	<0.001	0.803	0.849
0.789	<0.001	0.761	0.814

The p object created by `predictions()` includes the fitted values for each observation in the dataset, along with test statistics like p values and confidence intervals. p is a standard data frame, which means that we can manipulate it using the usual R or Python commands.

For example, we can check that the data frame includes 2825 and 11 columns:

```
dim(p)
```

```
[1] 2825   11
```

We can list the available column names:

```
colnames(p)
```

```
[1] "rowid"    "estimate"  "p.value"    "s.value"   "conf.low"   "conf.high"
[7] "df"       "outcome"   "incentive" "agecat"    "distance"
```

And we can extract individual columns and cells using the standard $ or [] syntaxes, or using data manipulation packages like `dplyr` or `data.table`:

```
p[1:4, "estimate"]
```

```
[1] 0.3652679 0.3540617 0.2653133 0.2997423
```

Users should be mindful of the fact that, by default, the p values held in this data frame correspond to a hypothesis test against a null of zero. In Section 5.5, we will see that it is easy to change this default null using the `hypothesis` argument.

5.2.2   Interesting grid

In many cases, analysts are not interested in model-based predictions for
each observation in their sample. Instead, they may prefer to build a grid of
predictor values that hold particular scientific or commercial interest.[4]

In `marginaleffects`, the main strategy to define custom grids is to use the
`newdata` argument and the `datagrid()` function. This function creates a
"typical" dataset with all variables held at their means or modes, except for
those we explicitly define.

```
datagrid(agecat = "18 to 35", incentive = c(0, 1), model = mod)

   distance    agecat incentive rowid
1 2.014541 18 to 35         0     1
2 2.014541 18 to 35         1     2
```

We can feed this `datagrid()` function to the `newdata` argument of
`predictions()`.[5]

```
predictions(mod,
  newdata = datagrid(agecat = "18 to 35", incentive = c(0, 1))
)
```

agecat	incentive	Estimate	Pr($>$|z|)	2.5 %	97.5 %
18 to 35	0	0.318	$<$0.001	0.279	0.361
18 to 35	1	0.780	$<$0.001	0.755	0.802

This shows that the estimated probability of seeking one's HIV status is about
32% for a participant who is between 18 and 35 years old, did not receive a
monetary incentive, and lives the average distance from the center.

One useful feature is that we can pass functions to `datagrid()`. These functions
will be applied to the named variables, and the output used to construct the
grid.

```
predictions(mod,
  newdata = datagrid(distance = 2, agecat = unique, incentive = max)
)
```

---

[4]Section 3.2.2

[5]When `datagrid()` is called as an argument to a `marginaleffects` function, we can omit
the `model` argument.

distance	agecat	incentive	Estimate	Pr($>$|z|)	2.5 %	97.5 %
2	<18	1	0.774	<0.001	0.725	0.816
2	18 to 35	1	0.780	<0.001	0.756	0.802
2	>35	1	0.815	<0.001	0.791	0.837

5.2.3   Representative grid

Sometimes, analysts do not want fine-grained control over the values of each predictor, but would rather compute predictions for some representative individual.[6] For example, we can compute a "Prediction at the Mean," that is, a prediction for a hypothetical representative individual whose personal characteristics are exactly average on numeric variables, and modal on categorical variables.

To achieve this, we can either call the `datagrid()` function without specifying the value of any variable, or we can use the `"mean"` shortcut.

```
p = predictions(mod, newdata = "mean")
p
```

Estimate	Pr($>$|z|)	2.5 %	97.5 %
0.78	<0.001	0.755	0.802

Our model expects that an individual with average or modal characteristics has a 78% probability of traveling to the test center to learn their HIV status.

Representative grids can be useful in some contexts, but they are not always the best choice. Sometimes there is simply no one in our sample who is exactly average on all relevant dimensions. When this average individual is fictional, the associated prediction may not be scientifically interesting or practically relevant.

5.2.4   Balanced grid

A common strategy in the analysis of experiments is to compute estimates on a "balanced grid."[7] This type of grid includes one row for each combination of unique values for the categorical (or binary) predictors, holding numeric variables at their means. To achieve this, we can either call `datagrid()` or use the `"balanced"` shortcut. These two calls are equivalent:

---

[6]Section 3.2.3

[7]Section 3.2.4

```
predictions(mod, newdata = datagrid(
  agecat = unique, incentive = unique, distance = mean
))

predictions(mod, newdata = "balanced")
```

agecat	incentive	distance	Estimate	$\mathrm{Pr}(>|z|)$	2.5 %	97.5 %
<18	0	2.01	0.311	<0.001	0.251	0.377
<18	1	2.01	0.773	<0.001	0.724	0.816
18 to 35	0	2.01	0.318	<0.001	0.279	0.361
18 to 35	1	2.01	0.780	<0.001	0.755	0.802
>35	0	2.01	0.367	<0.001	0.322	0.415
>35	1	2.01	0.815	<0.001	0.791	0.837

A balanced grid is often used with randomized experiments when the analyst wishes to give equal weight to each combination of treatment conditions in the calculation of marginal means.[8]

5.2.5   Counterfactual grid

Yet another set of predictor profiles to consider is the "counterfactual grid." The predictions made on such a grid allow us to answer questions such as:

What would the predicted outcomes be in our observed sample if everyone had received the treatment, or if everyone had received the control?

To create a counterfactual grid, we duplicate the full dataset, once for every value of the focal variable. For instance, if our dataset includes 2825 rows and we want to compute predictions for each value of the incentive variable (0 and 1), the counterfactual grid will include 5650 rows.

To make predictions on a counterfactual grid, we can call the datagrid() function with its grid_type="counterfactual" argument. Alternatively, we can use the variables argument to specify counterfactual values for the focal variable.

```
p = predictions(mod, variables = list(incentive = 0:1))
dim(p)
```

```
[1] 5650    12
```

---

[8]Section 5.3

These predictions are interesting, because they give us a first look at the kinds of counterfactual (potentially causal) queries that we will explore in Chapter 6. We can ask:

For each individual in the Thornton (2008) sample, what is the predicted probability of seeking information about HIV status in the counterfactual worlds where they receive a monetary incentive, and where they do not?

To answer this question, we rearrange the data and plot it.

```
library(ggplot2)

p = data.frame(
  Control = p[p$incentive == 0, "estimate"],
  Treatment = p[p$incentive == 1, "estimate"])

ggplot(p, aes(Control, Treatment)) +
  geom_abline(intercept = 0, slope = 1, linetype = "dashed") +
  geom_point() +
  labs(x = "Pr(outcome=1) when incentive = 0",
       y = "Pr(outcome=1) when incentive = 1") +
  xlim(0, 1) + ylim(0, 1) + coord_equal()
```

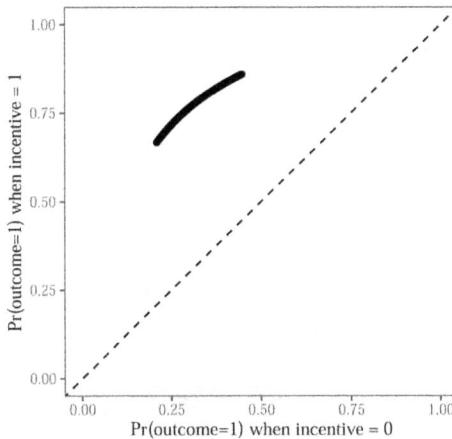

Figure 5.1: Predicted probabilities for counterfactual values of incentive.

On this graph, each point represents a single study participant.[9] The x-axis shows the predicted probability that *Outcome* equals 1 for an observed individual, after we artificially fix the `incentive` variable to 0. The y-axis shows the predicted probability that *Outcome* equals 1 for an individual with the same socio-demographic characteristics, after we artificially fix the `incentive`

---

[9]Since the points are tightly clustered, they appear as a curve at low resolution.

variable to 1. Each point thus shows what would be expected to happen if each individual did or did not receive the treatment.

Every point is well above the 45-degree line. This means that, for every observed combination of predictor values, for every participant in the study, our model says that changing the `incentive` variable from 0 to 1 increases the predicted probability that the person will seek to learn their HIV status.

5.3   Aggregation

Computing predictions for a large grid or for every observation in a dataset is useful, but the results can feel unwieldy. This section makes two principal arguments. First, it often makes sense to compute aggregated statistics, such as the average predicted outcome across the whole dataset, or by subgroups of the data. Second, the grid across which we aggregate can make a big difference to the results.

An "average prediction" is the outcome of a two step process. First, we compute predictions (fitted values) for each row in the original dataset. Then, we take the average of those predictions. This can be done manually by calling the `predictions()` function and taking the mean of estimates.

```
p = predictions(mod)
mean(p$estimate)
```

```
[1] 0.6916814
```

Alternatively, we can use the `avg_predictions()` function, which is a wrapper around `predictions()` that computes the average prediction directly.

```
avg_predictions(mod)
```

Estimate	Std. Error	z	Pr(>|z|)	2.5 %	97.5 %
0.692	0.00791	87.5	<0.001	0.676	0.707

The average predicted probability of seeking information about one's HIV status, across all the study participants in the Thornton (2008) sample, is about 69%.

Now, imagine that we want to know if the average predicted probability of `outcome=1` differs across age categories. To check this, we make the same function call, but add the `by` argument.

```
avg_predictions(mod, by = "agecat")
```

agecat	Estimate	Std. Error	z	Pr($>$|z|)	2.5 %	97.5 %
<18	0.670	0.0240	27.9	<0.001	0.623	0.717
18 to 35	0.673	0.0116	58.1	<0.001	0.650	0.695
>35	0.720	0.0121	59.5	<0.001	0.697	0.744

The average predicted probability of seeking one's test result is about 67% for minors and 72% for those above 35 years old. In Section 5.5 we will formally test if the difference between those two average predictions is statistically significant.

So far, we have taken averages over the empirical distribution of covariates, but analysts are not limited to that grid. One common alternative is to compute "marginal means" by averaging predictions across a balanced grid of predictors.[10] This is useful in experimental settings, when the observed sample is not representative of the population, and when we want to marginalize while giving equal weight to each treatment condition.

To compute marginal means, we call the same function, using the `newdata` and `by` arguments.

```
avg_predictions(mod, newdata = "balanced", by = "agecat")
```

agecat	Estimate	Std. Error	z	Pr($>$|z|)	2.5 %	97.5 %
<18	0.542	0.0261	20.7	<0.001	0.491	0.593
18 to 35	0.549	0.0135	40.6	<0.001	0.522	0.575
>35	0.591	0.0151	39.0	<0.001	0.561	0.621

Notice that the results are very different from the average predictions computed on the empirical grid. Now, the average predicted probability of seeking one's test result is estimated at 54% for minors and 59% for those above 35 years old.

The reason for this drastic change is that a balanced grid gives equal weight to each combination of categorical variables, while an empirical grid gives more weight to predictor values that are more frequent in the observed data. It turns out that, in the Thornton (2008) dataset, more participants belonged to the treatment than to the control group.

---

[10]This is the default approach taken by software packages like **emmeans** (Lenth, 2024).

```
table(dat$incentive)
```

```
  0    1
621 2204
```

Therefore, when we compute an average prediction on the empirical distribution, predicted outcomes in the `incentive=1` group are given more weight. This matters, because the average predicted probability that *Outcome* equals 1 is much higher in the treatment group than in the control group:

```
avg_predictions(mod, by = "incentive")
```

incentive	Estimate	Std. Error	z	Pr($>$|z|)	2.5 %	97.5 %
0	0.340	0.01890	18.0	<0.001	0.303	0.377
1	0.791	0.00862	91.7	<0.001	0.774	0.808

Thus, the group-wise averages for each age categories are smaller when computed over a balanced grid than when they are computed over the empirical distribution.

The last example of aggregation that we consider is done across a "counterfactual grid."[11] To build such a grid, every observation of the dataset is duplicated, and the `incentive` variable is fixed to different values in each counterfactual dataset. We can achieve this with the `variables` argument.

```
avg_predictions(
    mod,
    variables = list(incentive = c(0, 1)),
    by = "incentive")
```

incentive	Estimate	Std. Error	z	Pr($>$|z|)	2.5 %	97.5 %
0	0.339	0.01888	18.0	<0.001	0.302	0.376
1	0.791	0.00862	91.8	<0.001	0.774	0.808

This is equivalent to making unit-level predictions on two counterfactual grid, and then averaging those predictions manually.

```
p0 = predictions(mod, newdata = transform(dat, incentive = 0))
mean(p0$estimate)
```

```
[1] 0.3394248
```

---

[11]Section 3.2.5

```
p1 = predictions(mod, newdata = transform(dat, incentive = 1))
mean(p1$estimate)
```

```
[1] 0.7913343
```

5.4   Uncertainty

The `predictions()` family of functions accepts two arguments to control how to express the uncertainty around our estimates. First, the `conf_level` argument controls the size the confidence interval (default: 95%). Second, the vcov argument allows us to specify the type of standard errors to compute and report.

By default, `marginaleffects` functions return "classical" standard errors. These standard errors are computed using a procedure which assumes that errors are independently and identically distributed, and that rules out autocorrelation, clustering, or heteroskedasticity. Section 14.1.3 discusses these problems, and the solutions that statisticians have devised. For now, it suffices to point out that the vcov argument can be used to control the type of standard errors returned by the `predictions()` family of functions. For instance, setting vcov="HC3" will return heteroskedasticity-consistent standard errors.

```
avg_predictions(mod,
  by = "incentive",
  vcov = "HC3",
  conf_level = .9)
```

incentive	Estimate	Std. Error	z	Pr(>|z|)	5.0 %	95.0 %
0	0.340	0.01892	18.0	<0.001	0.309	0.371
1	0.791	0.00864	91.5	<0.001	0.777	0.805

In the Thornton dataset, we know that participants were recruited from different villages, and our dataset includes a `village` variable that records the place of origin of each participant. If the sampling or treatment assignment mechanism is related to these groups, it may be appropriate to report clustered standard errors (Abadie et al., 2022). To do this, we use the vcov argument and specify the clustering unit on the right-hand side of the ~ sign in a formula.

```
avg_predictions(mod,
  by = "incentive",
  vcov = ~ village,
  conf_level = .9)
```

incentive	Estimate	Std. Error	z	$\Pr(>|z|)$	5.0 %	95.0 %
0	0.340	0.0235	14.5	<0.001	0.301	0.378
1	0.791	0.0102	77.6	<0.001	0.774	0.808

The last class of uncertainty estimates that we consider here relies on re-sampling or simulation: bootstrap, simulation-based inference, or conformal inference. Chapter 14 devotes substantial space to explaining each of these approaches, but it is useful to note that these strategies are implemented using the inferences() function. To compute bootstrap confidence intervals, we can thus use the following command.

```
avg_predictions(mod,
  by = "incentive",
  conf_level = .9) |>
inferences(method = "boot", R = 1000)
```

incentive	Estimate	5.0 %	95.0 %
0	0.340	0.308	0.370
1	0.791	0.777	0.805

Notice that the intervals reported above are all slightly different, but still remain in the same ballpark.

5.5   Test

In the previous sections, we computed average predictions by age subgroups, and noted that there appeared to be differences in the likelihood that younger and older people would seek their HIV test results. That observation was based solely on the point estimates of the average predictions, and did not rely on a statistical test. Now, we consider how analysts can compare predictions more formally.

5.5.1   Null hypothesis tests

To begin, we compute the average predicted outcome for each age subgroup.

```
p = avg_predictions(mod, by = "agecat")
p
```

agecat	Estimate	Std. Error	z	Pr(>|z|)	2.5 %	97.5 %
<18	0.670	0.0240	27.9	<0.001	0.623	0.717
18 to 35	0.673	0.0116	58.1	<0.001	0.650	0.695
>35	0.720	0.0121	59.5	<0.001	0.697	0.744

The average predicted outcome is 67% for young adults and 72% for participants above 35 years old. The difference between these two averages is

```
p$estimate[3] - p$estimate[2]
```

```
[1] 0.04782993
```

To see if this risk difference is statistically significant, we use the `hypothesis` argument, as we did in Chapter 4.

```
p = avg_predictions(mod, by = "agecat",
   hypothesis = "b3 - b2 = 0")
p
```

Hypothesis	Estimate	Std. Error	z	Pr(>|z|)	2.5 %	97.5 %
b3-b2=0	0.0478	0.0167	2.86	0.00428	0.015	0.0806

The estimated difference between the 3rd and 2nd groups is about 5 percentage points, and the p value associated with this difference is 0.004. This crosses the conventional (but arbitrary) statistical significance threshold of $p = 0.05$. Many analysts would thus reject the null hypothesis that the average predicted probability of seeking one's HIV test results is the same in the 18 to 35 and >35 groups.

In many cases, we would like to make more than one comparison at a time. For instance, imagine an experiment with three groups: placebo, low dose, high dose. In that kind of study, we may want to compare each treatment group to the baseline or "reference" placebo group. Alternatively, we may want to do "sequential" comparisons, comparing each group to the one that directly precedes it. This can be helpful when there is a clear logical order between categories, and when we are interested in the progression across groups.

In `marginaleffects`, this can be done easily by supplying a formula to the
`hypothesis` argument. On the left-hand side, we set the comparison function:
difference, ratio, etc. On the right-hand side, we specify which estimates to
compare to one another: sequential, reference, etc.

```
avg_predictions(mod,
  by = "agecat",
  hypothesis = difference ~ sequential)
```

Hypothesis	Estimate	Std. Error	z	Pr(>|z|)	2.5 %	97.5 %
(18 to 35) − (<18)	0.00274	0.0267	0.103	0.91810	-0.0495	0.0550
(>35) - (18 to 35)	0.04783	0.0167	2.856	0.00428	0.0150	0.0806

By modifying the formula, we can compare each group to the "reference" or
baseline category, that is, participants under 18 years old.

```
avg_predictions(mod,
  by = "agecat",
  hypothesis = difference ~ reference)
```

Hypothesis	Estimate	Std. Error	z	Pr(>|z|)	2.5 %	97.5 %
(18 to 35) - (<18)	0.00274	0.0267	0.103	0.9181	−0.04952	0.055
(>35) - (<18)	0.05057	0.0269	1.880	0.0601	−0.00215	0.103

The `hypothesis` argument also allows us to conduct hypothesis tests by
subgroups. For example, consider this command, which computes average
predicted outcomes for each observed combination of `incentive` and `agecat`.

```
avg_predictions(mod,
    by = c("incentive", "agecat"))
```

incentive	agecat	Estimate	Std. Error	z	Pr(>|z|)	2.5 %	97.5 %
0	<18	0.312	0.0321	9.72	<0.001	0.249	0.375
0	18 to 35	0.324	0.0209	15.46	<0.001	0.283	0.365
0	>35	0.370	0.0235	15.74	<0.001	0.324	0.416
1	<18	0.771	0.0234	32.99	<0.001	0.725	0.816
1	18 to 35	0.778	0.0119	65.43	<0.001	0.755	0.801
1	>35	0.811	0.0117	69.04	<0.001	0.788	0.834

We can use the `hypothesis` argument in similar fashion as before, but add a vertical bar to specify that we want to compute sequential risk differences within subgroups.

```
avg_predictions(mod,
    by = c("incentive", "agecat"),
    hypothesis = difference ~ sequential | incentive)
```

incentive	Hypothesis	Estimate	Std. Error	z	Pr(>|z|)	2.5 %	97.5 %
0	(18 to 35) - (<18)	0.01111	0.0311	0.357	0.7213	-0.04994	0.0722
0	(>35) - (18 to 35)	0.04605	0.0216	2.131	0.0331	0.00369	0.0884
1	(18 to 35) - (<18)	0.00717	0.0254	0.283	0.7775	-0.04258	0.0569
1	(>35) - (18 to 35)	0.03330	0.0154	2.156	0.0311	0.00302	0.0636

This shows that, in the control group (`incentive=0`), the difference between the average predicted outcome for participants over 35 and for those between 18 and 35 is about 5 percentage points. However, in the treatment group (`incentive=1`), this difference is about 3 percentage points. Both of these differences are associated with relatively large Z statistics, and are thus statistically distinguishable from zero.

5.5.2 Equivalence tests

Flipping the logic around, the analyst could run an equivalence test to determine if the difference between average predicted outcomes in two subgroups is small enough to be considered negligible.[12] Imagine that, for domain-specific reasons, a risk difference smaller than 10 percentage points is considered "uninteresting," "negligible," or "equivalent to zero."

We start with this code, which computes the average predicted outcome for each age category.

```
avg_predictions(mod, by = "agecat")
```

agecat	Estimate	Std. Error	z	Pr(>|z|)	2.5 %	97.5 %
<18	0.670	0.0240	27.9	<0.001	0.623	0.717
18 to 35	0.673	0.0116	58.1	<0.001	0.650	0.695
>35	0.720	0.0121	59.5	<0.001	0.697	0.744

Then, we conduct a hypothesis test to compare the average predicted outcome for participants over 35 and for those under 18.

---

[12]Section 4.2

```
avg_predictions(mod,
    by = "agecat",
    hypothesis = "b3 - b1 = 0")
```

Hypothesis	Estimate	Std. Error	z	Pr(>|z|)	2.5 %	97.5 %
b3-b1=0	0.0506	0.0269	1.88	0.0601	-0.00215	0.103

Finally, we add the `equivalence` argument to specify the interval of practical equivalence.

```
avg_predictions(mod,
    by = "agecat",
    hypothesis = "b3 - b1 = 0",
    equivalence = c(-0.1, 0.1))
```

Hypothesis	Estimate	Std. Error	z	Pr(>|z|)	2.5 %	97.5 %	p (NonInf)	p (NonSup)	p (Equiv)
b3-b1=0	0.0506	0.0269	1.88	0.0601	-0.00215	0.103	<0.001	0.0331	0.0331

The p value associated with this test of equivalence is small. This suggests that we can reject the null hypothesis that the difference between average predicted outcomes is large or meaningful.

5.6   Visualization

In many cases, data analysts will want to visualize (potentially aggregated) predictions rather than report raw numeric values. This is easy to do with the `plot_predictions()` function, which has a syntax that closely parallels that of the other `marginaleffects` functions.

5.6.1   Unit predictions

As discussed in Chapter 3, the quantities derived from statistical models—predictions, counterfactual comparisons, and slopes—are typically conditional, in the sense that they depend on the values of all covariates in the model. This implies that each unit in our sample will be associated with its own prediction or effect estimate. In *Avoiding One-Number Summaries*, Harrell (2021) argues that data analysts should avoid the temptation to summarize individual-level estimates. Rather, Harrell argues, they should display the full distribution of estimates to convey a sense of the heterogeneity in our quantity of interest, across different combinations of predictor values.

Histograms and Empirical Cumulative Distribution Function (ECDF) plots are two common ways to visualize such a distribution. Since the output generated by the `predictions()` function is a standard data frame, it is easy to feed that object to any plotting function in R or Python, in order to craft good-looking visualizations. Here, we draw two plots, and combine them using the special + operator supplied by the `patchwork` package (Pedersen, 2024).

```
library(patchwork)

p = predictions(mod)

# Histogram
p1 = ggplot(p) +
  geom_histogram(aes(estimate, fill = factor(incentive))) +
  labs(x = "Pr(outcome = 1)", y = "Count", fill = "Incentive") +
  scale_fill_grey()

# Empirical Cumulative Distribution Function
p2 = ggplot(p) +
  stat_ecdf(aes(estimate, colour = factor(incentive))) +
  labs(x = "Pr(outcome = 1)",
       y = "Cumulative Probability",
       colour = "Incentive") +
  scale_colour_grey()

# Combined plots
p1 + p2
```

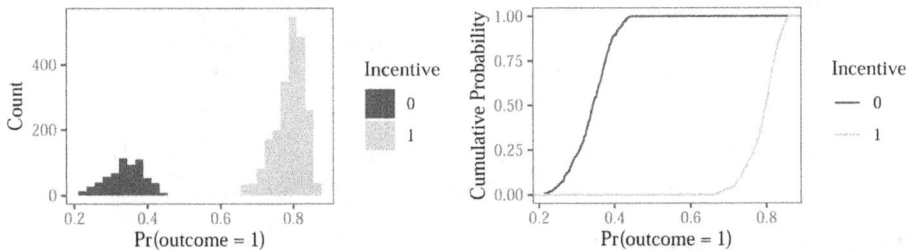

Figure 5.2: Distribution of unit-level predictions (fitted values), by treatment group.

The left side of Figure 5.2 is a histogram showing the distribution of predicted probabilities for each individual in the observed dataset. As usual, the x-axis represents the range of predicted outcomes, while the y-axis shows the number of study participants in each bin. By assigning different colors to the bins based on the treatment arm (`incentive` equal 0 or 1), we highlight one key feature of the distribution: predicted outcomes for people in the treatment group tend to be much higher than predicted outcomes for people in the control group. Indeed, the distribution of `outcome` probabilities without an incentive (black) is concentrated between 0.2 and 0.4, indicating a low probability of traveling to the test center. In contrast, the distribution of predicted `outcome` for participants who received a monetary incentive (grey) is concentrated between 0.70 and 0.85. This suggests that those who received an incentive are considerably more likely to seek their test results.

The right side of Figure 5.2 presents an ECDF plot. Again, the x-axis represents the range of predicted outcomes. This time, however, the y-axis indicates the empirical cumulative distribution, that is, the proportion of data points that are less than or equal to a specific value. For any given value on the x-axis, the height of the curve indicates the proportion of data points that are less than or equal to that value. For example, at 0.3 on the x-axis, we see that the `incentive=0` line is close to 0.25. This means that about 25% of participants in our sample have predicted `outcome` smaller than 0.3. Where the ECDF curve is steep, we know that much of our data is concentrated in that part of the distribution. With this in mind, we see that many of our predicted `outcome` values are clustered near 0.3 in the control group, and near 0.8 in the treatment group.

5.6.2   Marginal predictions

So far, we have focused on the full distribution of unit-level predictions. Often, we want to marginalize those unit-level estimates to plot average estimates instead. The first approach to do this uses the `by` argument.

When we use this argument, the function will (1) compute predictions for each observation in the actually observed dataset and (2) average unit-level predictions across some variable(s) of interest. This is equivalent to plotting the results of calling `avg_predictions()` using the `by` argument.

For example, if we want to compute the average predicted probability that `outcome` equals 1, by subgroup, we execute this code.

```
avg_predictions(mod, by = "incentive")
```

incentive	Estimate	Std. Error	z	Pr($>$|z|)	2.5 %	97.5 %
0	0.340	0.01890	18.0	<0.001	0.303	0.377
1	0.791	0.00862	91.7	<0.001	0.774	0.808

We plot the same results using the `plot_predictions()` function.

```
p1 = plot_predictions(mod, by = "incentive")
p2 = plot_predictions(mod, by = c("incentive", "agecat"))
p1 + p2
```

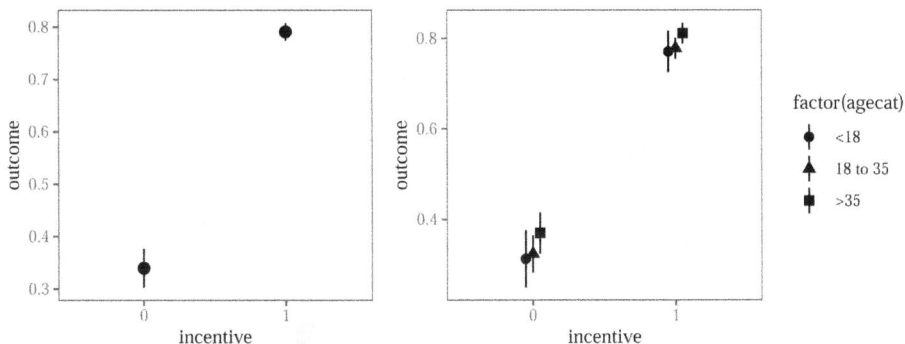

Figure 5.3: Marginal predicted probabilities that `outcome` equals 1.

These plots show that there is a very large gap between the predicted `outcome` for members of different `incentive` groups. In contrast, differences between age categories seem much more modest.

Note that the `plot_predictions()` function also accepts a `newdata` argument. This means that we can, for example, plot marginal means constructed by averaging across a balanced grid of predictors.[13]

```
plot_predictions(mod, by = "incentive", newdata = "balanced")
```

5.6.3 Conditional predictions

In some contexts, plotting marginal predictions may not be appropriate. For instance, when one of the predictors of interest is continuous, where there are many predictors, or much heterogeneity, the commands presented in the previous section may generate jagged plots that are difficult to read. In such cases, it can be useful to plot conditional predictions instead. In this context, the word "conditional" means that we are computing predictions, conditional on the values of the predictors in a constructed grid of representative values. However, unlike in the previous section, we do not average over several predictions before displaying the estimates. We fix the grid and display the predictions made for that grid immediately.

The `condition` argument of the `plot_predictions()` function does just that: build a grid of representative predictor values, compute predictions for each

---

[13]The plot is omitted because, in this particular case, it looks very similar to the one in Figure 5.3.

combination of predictor values, and plot the results. In the following examples, we fix one or more predictor to its unique values (categorical variables) or to an equally spaced grid from minimum to maximum (numeric variables). The other predictors in the model are held to their means or modes.

```
p1 = plot_predictions(mod,
  condition = "distance"
)
p2 = plot_predictions(mod,
  condition = c("distance", "incentive")
)
p3 = plot_predictions(mod,
  condition = c("distance", "incentive", "agecat")
)
(p1 + p2) / p3
```

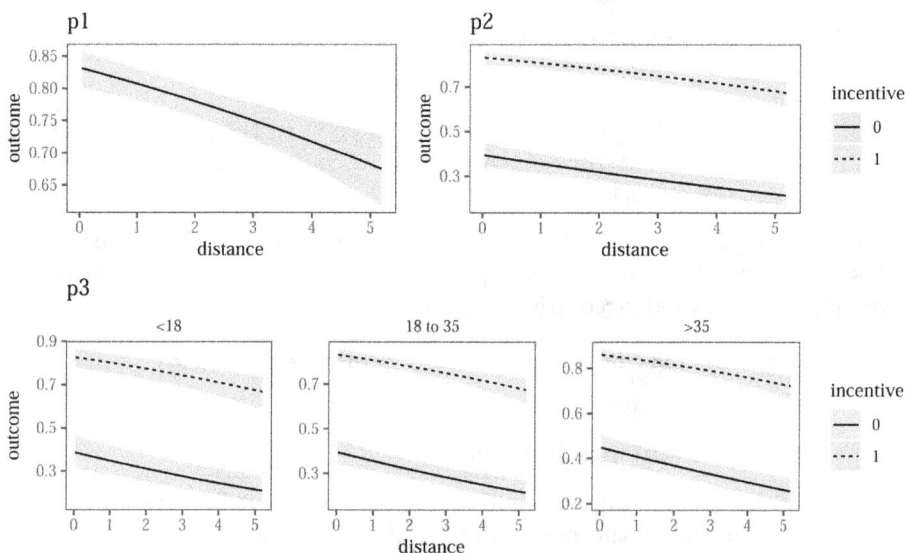

Figure 5.4: Predicted probability that `outcome` equals 1, conditional on incentive, age category, and distance. Other variables are held at their means or modes.

In each of these plots, the predicted probability that `outcome` equals one is lower for individuals who live far from the test center (right side of the plot). Moreover, we see that the dashed line are far above the solid lines, which tells us that the predicted outcome is higher for individuals who received a monetary incentive.

One useful feature of the `plot_predictions()` function is that it allows us to set the value of some variables explicitly in the `condition` argument. For example, to plot the predicted `outcome` for an individual above 35 years old, who did not receive a monetary incentive, for different values of distance:

```
plot_predictions(mod, condition = list(
    "distance", "agecat" = ">35", "incentive" = 0
))
```

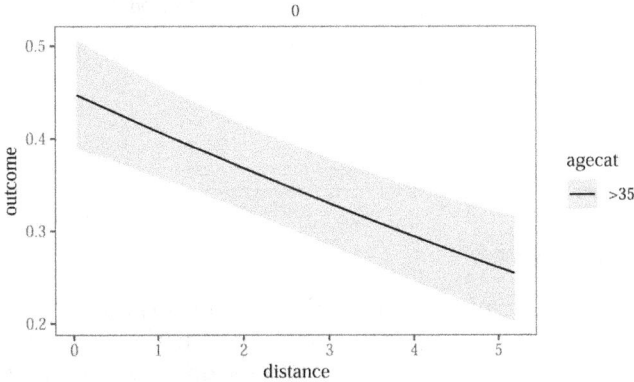

5.6.4 Customization

In R, the output of `plot_predictions()` is a standard `ggplot2` object. In Python, the object is compatible with the `plotnine` library. This makes it very easy for users to customize the appearance of their plots. For example, we can change the style theme and add a rug plot like so.

```
plot_predictions(mod, condition = "distance", rug = TRUE) +
    theme_grey() +
    ylim(c(.65, .81)) + xlim(c(2, 4.5))
```

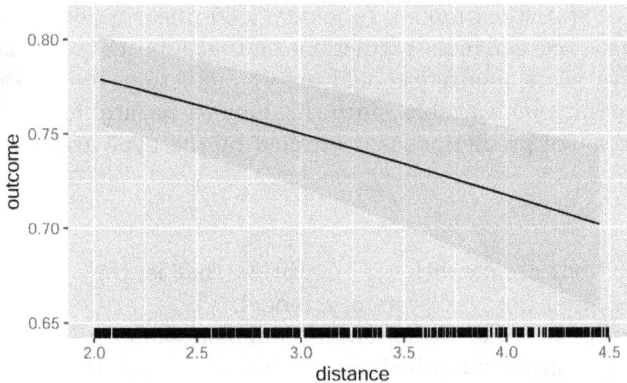

A more powerful but less convenient way to customize plots is to call the `draw=FALSE` argument. This will return a data frame with the raw values used

to create plots. You can then use these data to create your own plots with
base R graphics, `ggplot2`, or any other plotting functions you like.

```
plot_predictions(mod, by = "incentive", draw = FALSE)
```

```
  incentive  estimate    std.error statistic      p.value  s.value  conf.low
1         0 0.3397746 0.018899196  17.97825 2.884284e-72 237.6506 0.3027328
2         1 0.7908348 0.008620357  91.74038 0.000000e+00      Inf 0.7739393
  conf.high  df
1 0.3768163 Inf
2 0.8077304 Inf
```

5.7 Summary

This chapter defined a "prediction" as the outcome expected by a fitted model
for a given combination of predictor values. A prediction is a useful descriptive
quantity. It is an expectation or best guess for different individuals, units, or
subgroups of the population of interest.

The `predictions()` function from the `marginaleffects` package computes
predictions for a wide range of models. `avg_predictions()` aggregates unit-
level predictions. `plot_predictions()` displays predictions visually. The main
arguments of these functions are summarized in Table 5.1.

To clearly define predictions and attendant tests, analysts must make five
decisions.

First, the *Quantity*.

- Predictions can be computed on different scales, depending on the
  type of statistical model. In generalized linear models (GLM), for
  example, one can make predictions on the "link" or "response" scales.
  In most cases, analysts should report predictions on the same scale
  as the outcome variable, since this is most natural for readers.
- The scale of predictions is controlled by the `type` argument.

Second, the *Predictors*.

- Predictions are conditional quantities, that is, they depend on the
  values of all the predictors in a model.
- Analysts can make predictions for different combinations of predictor
  values, or grids: empirical, interesting, representative, balanced, or
  counterfactual.
- The predictor grid is defined by the `newdata` argument and the
  `datagrid()` function.

Table 5.1: Main arguments of the `predictions()`, `avg_predictions()`, and `plot_predictions()` functions.

Argument	
`model`	Fitted model used to make predictions.
`newdata`	Grid of predictor values.
`variables`	Make predictions on a counterfactual grid.
`vcov`	Standard errors: Classical, robust, clustered, etc.
`conf_level`	Size of the confidence intervals.
`type`	Type of predictions: response, link, etc.
`by`	Grouping variable for average predictions.
`byfun`	Custom aggregation functions.
`wts`	Weights used to compute average predictions.
`transform`	Post-hoc transformations.
`hypothesis`	Compare predictions to one another, conduct linear or non-linear hypothesis tests, or specify null hypotheses.
`equivalence`	Equivalence tests
`df`	Degrees of freedom for hypothesis tests.
`numderiv`	Algorithm used to compute numerical derivatives.
`...`	Additional arguments are passed to the `predict()` method supplied by the modeling package.

Third, the *Aggregation*.

- To simplify the presentation of results, it often makes sense to report average predictions. Analysts can choose between different aggregation schemes:
  − Unit-level predictions (no aggregation)
  − Average predictions
  − Average predictions by subgroup
  − Weighted average of predictions
- Predictions can be aggregated using the `avg_predictions()` function and the `by` argument.

Fourth, the *Uncertainty*.

- In `marginaleffects`, the `vcov` argument allows analysts to report classical, robust, or clustered standard errors around predictions.

- The `inferences()` function can compute uncertainty intervals via bootstrapping, simulation-based inference, or conformal prediction.

Fifth, the *Test*.

- A null hypothesis test aims to determine if a prediction (or a function of predictions) is different from a null hypothesis value. For example, an analyst may wish to check if two predictions are different from one another. Null hypothesis tests can be conducted using the `hypothesis` argument.
- An equivalence test aims to determine if a prediction (or a function of predictions) is similar to a reference value. Equivalence tests can be conducted using the `equivalence` argument.

6

Counterfactual comparisons

The main claim of this book is that the parameters of a statistical model are complex quantities that do not always have a straightforward meaning. Instead of trying to interpret them directly, we should treat parameters as a "resting stone on the way to prediction."

In this chapter, we go further and show that predictions can themselves act as a springboard for further analysis. We will see how to combine predictions into *counterfactual comparisons* that quantify the strength of association between two variables, or the effect of a cause.

Counterfactual thinking is fundamental to scientific inquiry and data analysis. Indeed, many of our most important research questions can be expressed as comparisons between hypothetical worlds.

- Would survival be more likely if patients received a new medication rather than a placebo?
- Would standardized test scores be higher if class sizes were smaller?
- Does participating in a micro-finance program increase household income?
- Do conservation policies improve forest coverage?

To answer questions like these, researchers must conduct thought experiments. They must make counterfactual predictions, and then compare those predictions.

Say we want to estimate the effect of conservation policies on forest coverage. To start, we ask: what is the forest coverage in a region without conservation policies? Then, we ask: what *would* be the forest coverage in the same region, if conservation policies were implemented? Finally, we compare the two quantities: what is the difference between the expected forest coverage in counterfactual worlds with and without conservation policies?

In that spirit, we can define a broad class of quantities of interest:

A counterfactual comparison is a function of two or more model-based predictions, made for different predictor values.

DOI: 10.1201/9781003560333-6

This definition is intimately linked to the theories of causal inference surveyed in Section 2.1.3. As we will see, a natural way to estimate the effect of an intervention is to use a statistical model to make predictions in two counterfactual worlds and to compare those predictions. When the conditions for causal identification are satisfied, this counterfactual comparison can be interpreted as a measure of the effect of X on Y.

But even where the conditions for causal identification are *not* satisfied, counterfactual comparisons remain very interesting statistical quantities. In that case, they can be treated as descriptive measures of the strength of association between two variables, holding other variables constant.

Sections 6.1 and 6.2 show that a vast array of estimands can be expressed as functions of two (or more) predictions: contrasts, risk differences, ratios, odds, lift, etc. Section 6.3 explains that we can aggregate counterfactual comparisons across different grids to compute quantities of interest like the average treatment effect (ATE) or the conditional average treatment effect (CATE). Section 6.4 discusses standard errors and confidence intervals, and Section 6.5 shows how we can contrast comparisons to one another, in view of exploring treatment effect heterogeneity. Section 6.6 concludes with some tips on data visualization.

6.1   Quantity

A counterfactual comparison is a function of two or more model-based predictions, made with different predictor values. To operationalize this quantity, we must make three decisions. First, what is the focal predictor whose effect on (or association with) the outcome we wish to estimate? Second, how does the focal predictor differ between counterfactual worlds? Third, what function do we use to compare predicted outcomes obtained for different values of the focal predictor?

To fix notation, consider a simple case where we fit a statistical model with outcome Y and focal predictor X, and use the parameter estimates to compute predictions. When the variable X is set to a specific value x, the model-based prediction is written $\hat{Y}_{X=x}$.

An analyst who is interested in model description, data description, or causal inference may want to estimate how the predicted outcome $\hat{Y}_X$ changes when we manipulate the predictor X. For example, how does the predicted outcome change when X increases by 1 unit, by one standard deviation, or when X changes from one specific value to another?

$$\hat{Y}_{X=x+1} - \hat{Y}_{X=x} \qquad \text{Increase of one unit}$$

$$\hat{Y}_{X=x+\sigma_X} - \hat{Y}_{X=x} \qquad \text{Increase of one standard deviation}$$

$$\hat{Y}_{X=max(X)} - \hat{Y}_{X=min(X)} \qquad \text{Increase from minimum to maximum}$$

$$\hat{Y}_{X=b} - \hat{Y}_{X=a} \qquad \text{Change between specific values } a \text{ and } b$$

In each of the examples above, we calculated the difference between two predicted outcomes, evaluated for different values of the focal predictor X. A simple difference is often the best starting point for interpretation, because it is simple and easy to grasp intuitively. But we are not restricted to this function.

$$\hat{Y}_{X=b} - \hat{Y}_{X=a} \qquad \text{Difference}$$

$$\frac{\hat{Y}_{X=b}}{\hat{Y}_{X=a}} \qquad \text{Ratio}$$

$$\frac{\hat{Y}_{X=b} - \hat{Y}_{X=a}}{\hat{Y}_{X=a}} \qquad \text{Lift}$$

In the special case where the predicted outcome $\hat{Y}_X$ is a probability, $\hat{Y}_{X=b} - \hat{Y}_{X=a}$ is called a risk difference, $\hat{Y}_{X=b}/\hat{Y}_{X=a}$ a risk ratio, and $\frac{\hat{Y}_{X=b}}{1-\hat{Y}_{X=b}} \Big/ \frac{\hat{Y}_{X=a}}{1-\hat{Y}_{X=a}}$ an odds ratio.

The rest of this chapter shows how to compute and interpret all of these quantities using the `marginaleffects` package.

For illustration, we fit a logistic regression model to data from Thornton (2008). The outcome is a binary variable which indicates if a study participant sought to learn their HIV status. The randomized treatment is a binary variable indicating whether the participant received a monetary incentive. To make the specification more flexible and improve precision, we interact the incentive indicator with two other predictors: a participant's age category and their distance from the test center.

```
library(marginaleffects)
dat = get_dataset("thornton")
mod = glm(outcome ~ incentive * (agecat + distance),
    data = dat, family = binomial)
summary(mod)
```

```
Coefficients:
                          Estimate Std. Error z value Pr(>|z|)
(Intercept)               -0.48617    0.29621  -1.641   0.1007
incentive                  2.05760    0.34421   5.978   <0.001
agecat18 to 35             0.07937    0.28872   0.275   0.7834
agecat>35                  0.34522    0.29467   1.172   0.2414
distance                  -0.18440    0.07236  -2.548   0.0108
incentive:agecat18 to 35  -0.05850    0.33500  -0.175   0.8614
incentive:agecat>35       -0.12468    0.34242  -0.364   0.7158
incentive:distance         0.02304    0.08256   0.279   0.7802
```

This is a simple logistic regression model. Yet, because the model includes interactions and a non-linear link function, interpreting raw model coefficients is difficult for most analysts, and essentially impossible for lay people.

Counterfactual comparisons are a compelling alternative to coefficient estimates, with many advantages for interpretation. First, counterfactual comparisons can be expressed directly on the scale of the outcome variable, rather than as complex functions like log-odds ratios. Second, counterfactual comparisons map directly onto what many people have in mind when they think of the effect of a treatment: what change can we expect in the outcome when a predictor changes? Finally, as the next sections show, the `marginaleffects` package makes it trivial to compute counterfactual comparisons. Data analysts can embrace the same workflow in model-agnostic fashion, applying similar post-estimation steps regardless of the kind of model they chose to estimate.

6.1.1   First steps: risk difference with a binary treatment

To begin, let us consider a simple estimand: the risk difference associated with a change in binary treatment. Specifically, we will estimate the expected change in `outcome` when the `incentive` variable is manipulated to equal 1 instead of 0.

An important factor to consider, when estimating such a quantity, is that counterfactual comparisons are *conditional* quantities. Except in the simplest cases, comparisons will depend on the values of all the predictors in a model. Each individual in a dataset may be associated with a different counterfactual comparison. Therefore, whenever the analyst computes a counterfactual comparison, they must explicitly define the values of the focal predictor, but also the values of all other covariates in the model.

Section 6.2 explores different ways to define a grid of predictor profiles. For now, we shall focus on a single individual with arbitrary characteristics.

```
grid = data.frame(distance = 2, agecat = "18 to 35", incentive = 1)
grid

  distance   agecat incentive
1        2 18 to 35         1
```

Our goal is to estimate the risk difference for someone between the ages of 18 to 35, who lives a distance of 2 from the test center.

$$\hat{Y}_{i=1,d=2,a=18 \text{ to } 35} - \hat{Y}_{i=0,d=2,a=18 \text{ to } 35}$$

To compute this quantity, we must compare model-based predictions with and without the incentive, holding all other unit characteristics constant. Using simple base R commands, we manipulate the grid, make predictions, and compare those predictions.

```
# Counterfactual grids of predictor values
g_treatment = transform(grid, incentive = 1)
g_control = transform(grid, incentive = 0)

# Counterfactual predictions
p_treatment = predictions(mod, newdata = g_treatment)$estimate
p_control = predictions(mod, newdata = g_control)$estimate

# Counterfactual comparison
p_treatment - p_control
```

```
[1] 0.465402
```

The same estimate can be obtained more easily, along with standard errors and test statistics, using the comparisons() function from the marginaleffects package.

```
comparisons(mod, variables = "incentive", newdata = grid)
```

Estimate	Std. Error	z	Pr(>|z|)	2.5 %	97.5 %
0.465	0.0293	15.9	<0.001	0.408	0.523

Our model suggests that, for a participant who is between 18 and 35 years old and lives a distance of 2 from the test center, moving from the control group to the treatment group increases the predicted probability that outcome equals one by $0.465 \times 100 = 46.5$ percentage points.

This result is interesting, but a note of caution is warranted. In this chapter, we will interpret the counterfactual comparison as a measure of the effect of a change in a focal predictor on the outcome *predicted by a model*. This is both a claim about the fitted model's behavior,[1] and a descriptive claim about the estimated association between a focal predictor and the outcome.[2] To go further and give counterfactual comparisons a causal interpretation would

---

[1] Section 2.1.1
[2] Section 2.1.2

require us to make strong assumptions. Chapter 8 discusses these assumptions in some detail.

6.1.2 Comparison functions

So far, we have measured the effect of a change in predictor solely by looking at differences in predicted outcomes. Differences are typically the best starting point for interpretation, because they are simple and easy to grasp intuitively. Nevertheless, in some contexts it can make sense to use different functions to compare counterfactual predictions, such as ratios, lift, or odds ratios.

To compute the ratio of predicted outcomes associated to a change in incentive, we use the `comparison="ratio"` argument.

$$\frac{\hat{Y}_{i=1,d=2,a=18\text{-}35}}{\hat{Y}_{i=0,d=2,a=18\text{-}35}}$$

```
comparisons(mod,
  variables = "incentive",
  comparison = "ratio",
  hypothesis = 1,
  newdata = grid)
```

Estimate	Std. Error	z	Pr(>|z|)	2.5 %	97.5 %
2.48	0.211	7	<0.001	2.06	2.89

The predicted outcome is nearly 2.5 times as large for a participant in the treatment group, who is between 18 and 35 years old and lives at a distance of 2 from the test center, than for a participant in the control group with the same socio-demographic characteristics. Note that, in the code above, we set `hypothesis=1` to test against the null hypothesis that these two predicted probabilities are identical (ratio of 1). The standard error is small and the z statistic large.

Therefore, we can reject the null hypothesis that the predicted outcomes are the same in the treatment and control arms of this trial. We can reject the hypothesis that the ratio between the predicted probabilities that someone in the treatment group (`incentive=1`) and someone in the control group (`incentive=0`) will get their test results is 1.

To compute the lift, we would proceed in the same way, by setting
`comparison="lift"`.

$$\frac{\hat{Y}_{i=1,d=2,a=18\text{-}35} - \hat{Y}_{i=0,d=2,a=18\text{-}35}}{\hat{Y}_{i=0,d=2,a=18\text{-}35}}$$

```
comparisons(mod,
  variables = "incentive",
  comparison = "lift",
  newdata = grid)
```

Estimate	Std. Error	z	Pr(>|z|)	2.5 %	97.5 %
1.48	0.211	7	<0.001	1.06	1.89

Finally, it is useful to note that the `comparison` argument accepts arbitrary
functions. This is an extremely powerful feature, as it allows analysts to specify
fully customized comparisons between a `hi` prediction (e.g., treatment) and a
`lo` prediction (e.g., control). To illustrate, we compute a log odds ratio based
on average predictions.[3]

```
lnor = function(hi, lo) {
  log((mean(hi) / (1 - mean(hi))) / (mean(lo) / (1 - mean(lo))))
}
comparisons(mod,
  variables = "incentive",
  comparison = lnor)
```

Estimate	Std. Error	z	Pr(>|z|)	2.5 %	97.5 %
2	0.0991	20.2	<0.001	1.8	2.19

6.2 Predictors

The predictors in a model can be divided in two categories: focal and adjustment
variables. Focal variables are the key predictors in a counterfactual analysis.
They are the variables whose effect on (or association with) the outcome we
wish to quantify. In contrast, adjustment (or control) variables are incidental
to the principal analysis. They can be included in a model to increase its

---

[3]We take the log odds ratios of the averages, rather than the average log odds ratio,
because odds ratios are non-collapsible.

flexibility, improve fit, control for confounders, or to check if treatment effects vary across subgroups of the population. However, the effect of an adjustment variable is not of inherent interest in a counterfactual analysis.[4]

As noted above, counterfactual comparisons are conditional quantities, which means that they typically depend on the values of all the predictors in a model. Therefore, when computing a comparison, we must decide where to evaluate it in the predictor space. We must decide what values to assign to both the focal and adjustment variables.

6.2.1   Focal variables

When estimating counterfactual comparisons, our goal is to determine what happens to the predicted outcome when one or more focal predictors change. Obviously, the kind of change we are interested in depends on the nature of the focal predictors. Let us consider four common cases: binary, categorical, numeric, and cross-comparisons.

For pedagogical purposes, we will treat each of the predictors in our model as a focal variable in turn: `incentive`, `agecat`, and `distance`. Note, however, that only the `incentive` variable was randomized in the Thornton (2008) study. In most real-world applications, there will only be one or two focal predictors per statistical model.[5]

6.2.1.1   Change in binary predictors

By default, when the focal predictor is binary, `marginaleffects` returns the value of the difference in predicted outcome associated to a change from the control (0) to the treatment (1) group.

```
comparisons(mod, variables = "incentive", newdata = grid)
```

Estimate	Std. Error	z	Pr(>|z|)	2.5 %	97.5 %
0.465	0.0293	15.9	<0.001	0.408	0.523

If an analyst wants to compute the effect of a change in the opposite direction, they can specify that change explicitly using the list syntax.

```
comparisons(mod, variables = list("incentive" = c(1, 0)), newdata = grid)
```

---

[4]Interpreting the parameters associated with adjustment variables as measures of association or effect is generally not recommended. See the discussion of Table 2 fallacy in Section 2.2 and Westreich and Greenland (2013).

[5]See Section 2.1.3 for a discussion of the problems that can arise when multiple focal variables are included in a single model. Also see Goldsmith-Pinkham et al. (Forthcoming).

Estimate	Std. Error	z	Pr(>|z|)	2.5 %	97.5 %
−0.465	0.0293	−15.9	<0.001	−0.523	−0.408

Moving from the treatment to the control group on the `incentive` variable is a associated with a change of −47 percentage points in the predicted probability that `outcome` equals 1.

6.2.1.2 Change in categorical predictors

The same approach can be used when we are interested in changes in a categorical variable with multiple levels. For example, if we want to know how changes in the `agecat` variable affect the predicted probability that `outcome` equals 1, we use the `variables` argument.

```
comparisons(mod, variables = "agecat", newdata = grid)
```

Contrast	Estimate	Std. Error	z	Pr(>|z|)	2.5 %	97.5 %
18 to 35 − <18	0.00359	0.0294	0.122	0.903	−0.0540	0.0611
>35 − <18	0.03587	0.0294	1.221	0.222	−0.0217	0.0935

Moving from the *<18* category to the *18 to 35* category increases the predicted probability that `outcome` equals 1 by 0.4 percentage points. Moving from the *<18* to the *>35* age bracket increases the predicted probability by 3.6 percentage points.

By default, the `comparisons()` function returns comparisons between every level of the categorical predictor and its reference level, or first category. We can modify the `variables` argument to compare specific categories, or to compare all categories to its preceding level, sequentially: *<18* to *18 to 35*, and *18 to 35* to *>35*.

```
# Specific comparison
comparisons(mod,
  variables = list("agecat" = c("18 to 35", ">35")),
  newdata = grid)

# Sequential comparisons
comparisons(mod,
  variables = list("agecat" = "sequential"),
  newdata = grid)
```

6.2.1.3   Change in numeric predictors

When the focal predictor is numeric, we can once again use the `variables` argument.

```
comparisons(mod, variables = "distance", newdata = grid)
```

Estimate	Std. Error	z	Pr(>|z|)	2.5 %	97.5 %
−0.0289	0.00742	−3.89	<0.001	−0.0434	−0.0143

This table shows the effect of increasing `distance` by 1 unit on the predicted value of `outcome`. Importantly, this corresponds to an increase of 1 unit from the value of `distance` encoded in the predictor grid.

```
grid
```

```
   distance    agecat incentive
1         2 18 to 35         1
```

For a person between the ages of 18 and 35, in the treatment group, moving from 2 to 3 units on the `distance` variable is associated with a change of −2.9 percentage points on the predicted probability of the outcome.

One may be interested in different magnitudes of change in the focal predictor `distance`. For example, the effect of a 5 unit (or 1 standard deviation) increase in `distance` on the predicted value of the outcome. Alternatively, the analyst may want to assess the effect of a change between two specific values of `distance`, or across the interquartile (or full) range of the data. All of these options are easy to implement using the `variables` argument.

```
# Increase of 5 units
comparisons(mod, variables = list("distance" = 5), newdata = grid)

# Increase of 1 standard deviation
comparisons(mod, variables = list("distance" = "sd"), newdata = grid)

# Change between specific values
comparisons(mod, variables = list("distance" = c(0, 3)), newdata = grid)

# Change across the interquartile range
comparisons(mod, variables = list("distance" = "iqr"), newdata = grid)

# Change across the full range
comparisons(mod, variables = list("distance" = "minmax"), newdata = grid)
```

6.2.1.4 Cross-comparisons

Sometimes, an analyst wants to assess the joint or combined effect of manipulating two predictors. In a medical study, for example, we may be interested in the change in survival rates for people who both receive a new treatment and make a dietary change. In our running example, it may be interesting to know how much the predicted probability of outcome would change if we modified both the distance and incentive variables simultaneously. To check this, we use the cross argument.

```
cmp = comparisons(mod,
  variables = c("incentive", "distance"),
  cross = TRUE,
  newdata = grid)
cmp
```

C: distance	C: incentive	Estimate	Std. Error	z	Pr(>|z|)	2.5 %	97.5 %
+1	1 − 0	0.437	0.0305	14.3	<0.001	0.377	0.496

These results show that a simultaneous increase of 1 unit in the distance variable and between 0 and 1 on the incentive variable is associated with a change of 0.437 in the predicted outcome.

6.2.2 Adjustment variables

In a typical counterfactual analysis, the researcher is not interested in a change in the adjustment variables themselves. Nevertheless, since the value of a counterfactual comparison depends on where it is evaluated in the predictor space, we must imperatively define the full grid of focal and adjustment variables.

Much like in Chapter 5, where we computed predictions for different profiles, we now estimate counterfactual comparisons on empirical, interesting, representative, and balanced grids.

6.2.2.1 Empirical distribution

By default, the comparisons() function returns estimates for every row of the original data frame that was used to fit the model. The Thornton (2008) dataset includes 2825 complete observations (after dropping missing data), so the next command will yield 2825 estimates.

```
comparisons(mod, variables = "incentive")
```

Estimate	Std. Error	z	Pr(>|z|)	2.5 %	97.5 %
0.458	0.0689	6.64	<0.001	0.323	0.593
0.463	0.0666	6.95	<0.001	0.332	0.593
0.488	0.0611	7.99	<0.001	0.368	0.608
0.482	0.0603	7.99	<0.001	0.364	0.600
0.471	0.0632	7.45	<0.001	0.347	0.595
		2815 rows omitted			
0.423	0.0370	11.43	<0.001	0.350	0.495
0.416	0.0393	10.59	<0.001	0.339	0.493
0.423	0.0370	11.43	<0.001	0.350	0.495
0.428	0.0355	12.05	<0.001	0.358	0.498
0.454	0.0377	12.02	<0.001	0.380	0.528

If we do not specify the `variables` argument, `comparisons()` computes distinct differences for all the variables. Here, there are 4 possible differences, so we get $4 \times 2825 = 11300$ rows.

```
cmp = comparisons(mod)
nrow(cmp)
```

```
[1] 11300
```

```
cmp
```

Term	Contrast	Estimate	Std. Error	z	Pr(>|z|)	2.5 %	97.5 %
agecat	18 to 35 − <18	0.0184	0.0665	0.277	0.782	−0.1120	0.149
agecat	18 to 35 − <18	0.0181	0.0655	0.277	0.782	−0.1102	0.147
agecat	18 to 35 − <18	0.0151	0.0544	0.278	0.781	−0.0916	0.122
agecat	18 to 35 − <18	0.0165	0.0593	0.278	0.781	−0.0998	0.133
agecat	18 to 35 − <18	0.0176	0.0634	0.277	0.782	−0.1067	0.142
		11290 rows omitted					
incentive	1 − 0	0.4227	0.0370	11.431	<0.001	0.3502	0.495
incentive	1 − 0	0.4161	0.0393	10.585	<0.001	0.3391	0.493
incentive	1 − 0	0.4227	0.0370	11.427	<0.001	0.3502	0.495
incentive	1 − 0	0.4280	0.0355	12.049	<0.001	0.3584	0.498
incentive	1 − 0	0.4539	0.0377	12.024	<0.001	0.3799	0.528

Since the output of `comparisons()` is a simple data frame, we can easily plot the full distribution of unit-specific risk differences.

```
library(ggplot2)

ggplot(cmp, aes(x = estimate)) +
  geom_histogram(bins = 30) +
  facet_grid(. ~ term + contrast, scales = "free") +
  labs(x = "Estimated change in predicted outcome", y = "Count")
```

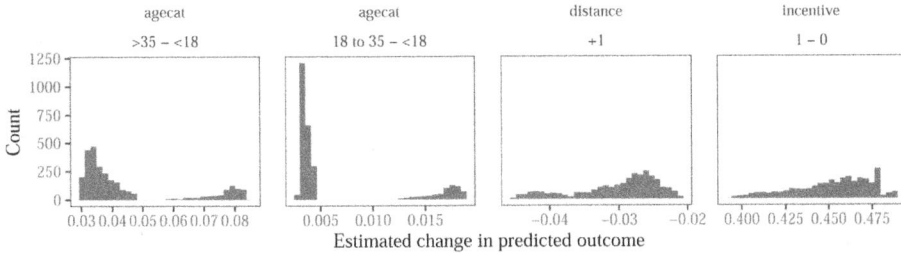

Figure 6.1: Distribution of unit-level risk differences associated with changes in each of the predictors.

The x-axes in Figure 6.1 show the estimated effects of changes in the predictors on the predicted value of the outcome. The y-axes show the prevalence of each estimate across the full sample.

There appears to be considerable heterogeneity. For example, consider the right-most panel, which plots the distribution of unit-level contrasts for the incentive variable. This panel shows that for some participants, the model predicts that moving from the control to the treatment condition would increase the predicted probability that outcome equals one by about 0.5 points. For others, the estimated effect of this change can be as low as 0.4.

6.2.2.2 Interesting grid

In some contexts, we want to estimate a contrast for a specific individual with characteristics of interest. To achieve this, we can supply a data frame to the newdata argument.

The code below shows the expected change in the predicted probability of the outcome associated with a change in incentive, for a few individuals with interesting characteristics. As in Chapter 5, we use the datagrid() function as a convenient mechanism to create a grid of profiles of interest.

```
comparisons(mod,
  variables = "incentive",
  newdata = datagrid(agecat = unique, distance = mean))
```

agecat	distance	Estimate	Std. Error	z	Pr(>|z|)	2.5 %	97.5 %
<18	2.01	0.479	0.0608	7.87	<0.001	0.360	0.598
18 to 35	2.01	0.466	0.0293	15.87	<0.001	0.408	0.523
>35	2.01	0.438	0.0342	12.79	<0.001	0.371	0.505

Notice that the estimated effects of `incentive` on the predicted probability that `outcome=1` differ depending on participants' age. Indeed, our model estimates that the difference between treatment and control would be 46.6 percentage points in the central age category, but 43.8 percentage points in the older age category.

6.2.2.3 Representative grids

A common alternative is to compute a comparison or risk difference "at the mean." The idea is to create a "representative" or "synthetic" profile for an individual whose characteristics are completely average or modal. Then, we report the comparison for this specific hypothetical individual. To do this, we use the "mean" shortcut in the `newdata` argument.

```
comparisons(mod, variables = "incentive", newdata = "mean")
```

Estimate	Std. Error	z	Pr(>|z|)	2.5 %	97.5 %
0.466	0.0293	15.9	<0.001	0.408	0.523

The main advantage of this approach is that it is fast and cheap from a computational standpoint. The disadvantage is that the interpretation is somewhat ambiguous. Often, the population includes no individual at all who is perfectly average across all dimensions and it is not always clear why we should be interested in such an individual. This matters because, in some cases, a "comparison at the mean" can differ significantly from an "average comparison."[6]

6.2.2.4 Balanced grids

Balanced grids, introduced in Section 3.2, include all unique combinations of categorical variables, while holding numeric variables at their means. This is particularly useful in experimental contexts, where the sample is not representative of the target population, and where we want to treat each combination of treatment conditions similarly.

---

[6]Section 6.3

```
cmp = comparisons(mod, variables = "incentive", newdata = "balanced")
as.data.frame(cmp)
```

	rowid	term	contrast	estimate	std.error	statistic	p.value
1	1	incentive	1 - 0	0.4788406	0.06084969	7.869237	3.568100e-15
2	2	incentive	1 - 0	0.4655786	0.02933085	15.873341	9.693996e-57
3	3	incentive	1 - 0	0.4379624	0.03423944	12.791165	1.836999e-37
4	4	incentive	1 - 0	0.4788406	0.06084969	7.869237	3.568100e-15
5	5	incentive	1 - 0	0.4655786	0.02933085	15.873341	9.693996e-57
6	6	incentive	1 - 0	0.4379624	0.03423944	12.791165	1.836999e-37

	s.value	conf.low	conf.high	predicted_lo	predicted_hi	predicted	agecat
1	47.99377	0.3595774	0.5981038	0.2978325	0.7766732	0.2978325	<18
2	186.07281	0.4080912	0.5230660	0.3146933	0.7802718	0.3146933	18 to 35
3	122.03399	0.3708543	0.5050704	0.3746268	0.8125892	0.3746268	>35
4	47.99377	0.3595774	0.5981038	0.2978325	0.7766732	0.7766732	<18
5	186.07281	0.4080912	0.5230660	0.3146933	0.7802718	0.7802718	18 to 35
6	122.03399	0.3708543	0.5050704	0.3746268	0.8125892	0.8125892	>35

	distance	outcome
1	2.014541	1
2	2.014541	1
3	2.014541	1
4	2.014541	1
5	2.014541	1
6	2.014541	1

The code above converted the results to a `data.frame`. This is because `marginaleffects` automatically hides some columns when printing output. Calling `as.data.frame()` ensures that all columns are printed, which makes it easier to see the structure of the balanced grid.

6.3 Aggregation

As discussed above, the default behavior of `comparisons()` is to estimate quantities of interest for all the actually observed units in our dataset, or for each row of the dataset supplied to the `newdata` argument. Sometimes, it is convenient to marginalize those conditional estimates to obtain an average (or marginal) estimates.

Several key quantities of interest can be expressed as average counterfactual comparisons. For example, when certain assumptions are satisfied, the average treatment effect (ATE) of X on Y can be defined as the expected difference between outcomes under treatment or control:

$$E[Y_{X=1} - Y_{X=0}],$$

where the expectation is taken over the distribution of adjustment variables. In Chapter 8, we will see that estimands like the Average Treatment Effect on the Treated (ATT) or Average Treatment Effect on the Untreated (ATU) can be defined analogously.

To compute an average counterfactual comparison, we proceed in four steps:

1. Compute predictions for every row of the dataset in the counterfactual world where all observations belong to the treatment condition.
2. Compute predictions for every row of the dataset in the counterfactual world where all observations belong to the control condition.
3. Take the differences between the two vectors of predictions.
4. Average the unit-level estimates across the whole dataset, or within subgroups.

Previously, we called the `comparisons()` function to compute unit-level counterfactual comparisons. To return average estimates, we call the same function, with the same arguments, but add the `avg_` prefix.

```
avg_comparisons(mod, variables = "incentive")
```

Estimate	Std. Error	z	Pr(>|z|)	2.5 %	97.5 %
0.452	0.0208	21.8	<0.001	0.411	0.493

On average, across all participants in the study, moving from the control to the treatment group is associated with a change of 45.2 percentage points in the predicted probability that `outcome` equals one. This result is equivalent to computing unit-level estimates and taking their mean.

```
cmp = comparisons(mod, variables = "incentive")
mean(cmp$estimate)
```

```
[1] 0.45192
```

Using the `by` argument, we can compute the average risk difference with respect to `incentive`, for each age subcategory.

```
avg_comparisons(mod, variables = "incentive", by = "agecat")
```

agecat	Estimate	Std. Error	z	Pr(>|z|)	2.5 %	97.5 %
<18	0.475	0.0605	7.85	<0.001	0.356	0.593
18 to 35	0.461	0.0290	15.89	<0.001	0.404	0.518
>35	0.435	0.0338	12.84	<0.001	0.368	0.501

On average, for participants in the <18 age bracket, moving from the control to the treatment group is associated with a change of 47.5 percentage points in the predicted probability that outcome equals one. This average risk difference is estimated at 43.5 percentage points for participants above 35 years old.

We can also compute an average risk difference for individuals with specific profiles, by specifying the grid of predictors using the newdata argument. For example, if we are interested in an average treatment effect that only applies to study participants who actually belonged to the treatment group, we can call:[7]

```
avg_comparisons(mod,
  variables = "incentive",
  newdata = subset(incentive == 1)
)
```

Estimate	Std. Error	z	Pr(>|z|)	2.5 %	97.5 %
0.452	0.0208	21.8	<0.001	0.411	0.493

6.3.1 Average predictions vs. average comparisons

Before moving on, it is useful to take a slight detour to highlight the relationships between three of the most important quantities of interest we have considered so far: average predictions, average counterfactual predictions, and average counterfactual comparisons.

To illustrate these concepts, we momentarily shift our focus to the Palmer Penguins dataset (Horst et al., 2020), which contains measurements of body mass and flipper lengths for three species of penguins: Adelie, Chinstrap, and Gentoo. On average, the dataset shows that Gentoo penguins are heaviest and have the longest flippers.

```
penguins = get_dataset("penguins", "palmerpenguins")

aggregate(
  cbind(body_mass_g, flipper_length_mm) ~ species,
  FUN = mean,
  data = penguins)

    species body_mass_g flipper_length_mm
1    Adelie    3700.662          189.9536
2 Chinstrap    3733.088          195.8235
3    Gentoo    5076.016          217.1870
```

---

[7]In the current example, computing the average across the full empirical distribution or in the subset of actually treated units does not make much difference, which is often the case for randomized controlled experiments. As Chapter 8 shows, this is not the case in general.

We fit a linear regression model to predict flipper length based on body mass and species. Average predictions are calculated over the observed distribution of covariates within each subset of interest. If Gentoo penguins are heavier than Adelie, that difference in predictors will be reflected in the average predictions.

```
fit = lm(flipper_length_mm ~ body_mass_g * species, data = penguins)

avg_predictions(fit, by = "species")
```

species	Estimate	Std. Error	z	Pr($>$|z|)	2.5 %	97.5 %
Adelie	190	0.435	436	<0.001	189	191
Chinstrap	196	0.649	302	<0.001	195	197
Gentoo	217	0.482	450	<0.001	216	218

Now, consider what happens when we compute average *counterfactual* predictions instead. As explained in Section 5.2.5, this involves duplicating the entire dataset three times—once for each penguin species—and fixing the focal variable to be constant in the three dataset. Then, we call the `predict()` and `mean()` functions from base R, in order to return the average expected flipper length.

```
# Adelie
penguins |>
  transform(species = "Adelie") |>
  predict(fit, newdata = _) |>
  mean(na.rm = TRUE)
```

```
[1] 193.2994
```

```
# Chinstrap
penguins |>
  transform(species = "Chinstrap") |>
  predict(fit, newdata = _) |>
  mean(na.rm = TRUE)
```

```
[1] 201.403
```

```
# Gentoo
penguins |>
  transform(species = "Gentoo") |>
  predict(fit, newdata = _) |>
  mean(na.rm = TRUE)
```

```
[1] 209.2844
```

Instead of this manual calculation, we can compute the average counterfactual predictions with the `avg_predictions()` function and its `variables` and `by` arguments.

```
p = avg_predictions(fit, variables = "species", by = "species")
p
```

species	Estimate	Std. Error	z	Pr(>|z|)	2.5 %	97.5 %
Adelie	193	0.646	299	<0.001	192	195
Chinstrap	201	1.027	196	<0.001	199	203
Gentoo	209	0.968	216	<0.001	207	211

Note that since we have replicated the full datasets, the distribution of body mass is now identical in all three datasets. By construction, the average counterfactual predictions are *ceteris paribus* quantities, that allow us to compare species while holding covariates constant. As a result of "ignoring" body mass, the gaps between average counterfactual predictions are smaller than the gaps between average predictions.

Finally, average counterfactual comparisons measure the differences between these counterfactual predictions, giving us the estimated effect of species on flipper length, while controlling for body mass. Average counterfactual comparisons can be computed directly by taking the difference between average counterfactual predictions, or by calling `avg_comparisons()`.

```
diff(p$estimate)
```

```
[1] 8.103662 7.881389
```

```
avg_comparisons(fit, variables = list(species = "sequential"))
```

Contrast	Estimate	Std. Error	z	Pr(>|z|)	2.5 %	97.5 %
Chinstrap − Adelie	8.10	1.21	6.68	<0.001	5.73	10.5
Gentoo − Chinstrap	7.88	1.41	5.58	<0.001	5.11	10.6

This example illustrates how counterfactual predictions and comparisons allow us to conduct "all else equal" analyses—examining the relationship between variables while holding other factors constant. The differences between species are smaller in the counterfactual analysis than in the raw data because we have "controlled for" the fact that some species tend to be heavier than others.

In the next section, we go back to our running example.

6.4   Uncertainty

By default, the standard errors around contrasts are estimated using the delta method and the classical variance-covariance matrix supplied by the modeling software.[8] For many of the more common statistical models, we can rely on the `sandwich` or `clubSandwich` packages to report "robust" standard errors, confidence intervals, and p values (Zeileis et al., 2020; Pustejovsky, 2023). We can also use the `inferences()` function to compute bootstrap or simulation-based estimates of uncertainty.

Using the `modelsummary` package, we can report estimates with different uncertainty estimates in a single table, with models displayed side-by-side (Table 6.1).[9]

```
library(modelsummary)

models = list(
  "Heteroskedasticity" = avg_comparisons(mod, vcov = "HC3"),
  "Clustered" = avg_comparisons(mod, vcov = ~village),
  "Bootstrap" = avg_comparisons(mod) |> inferences("boot")
)

modelsummary(models,
  statistic = "conf.int", fmt = 4, gof_map = "nobs",
  shape = term + contrast ~ model
)
```

6.5   Test

Earlier in this chapter, we estimated average counterfactual comparisons for different age subgroups. Now, we will see how to conduct hypothesis tests on these quantities, in order to compare them to one another.

Imagine one is interested in the following question:

> *Does moving from `incentive=0` to `incentive=1` have a bigger effect on the predicted probability that `outcome=1` for older or younger participants?*

---

[8]Section 14.1

[9]The `statistic` argument allows us to display confidence intervals instead of the default standard errors. The `fmt` determines the number of digits to display. `gof_omit` omits all goodness-of-fit statistics from the bottom of the table. `shape` indicates that models should be displayed as columns and terms and contrasts as structured rows.

Table 6.1: Alternative ways to compute uncertainty about counterfactual comparisons.

		Heteroskedasticity	Clustered	Bootstrap
agecat	18 to 35 − <18	0.0065	0.0065	0.0065
		[−0.0457, 0.0587]	[−0.0468, 0.0599]	[−0.0470, 0.0580]
	>35 − <18	0.0449	0.0449	0.0449
		[−0.0079, 0.0977]	[−0.0033, 0.0931]	[−0.0109, 0.0977]
distance	+1	−0.0303	−0.0303	−0.0303
		[−0.0427, −0.0179]	[−0.0449, −0.0157]	[−0.0428, −0.0185]
incentive	1 − 0	0.4519	0.4519	0.4519
		[0.4110, 0.4928]	[0.4094, 0.4945]	[0.4127, 0.4915]
Num.Obs.		2825	2825	2825

To answer this, we can start by using the `avg_comparisons()` and its `by` argument to estimate sub-group specific average risk differences.

```
cmp = avg_comparisons(mod,
  variables = "incentive",
  by = "agecat")
cmp
```

agecat	Estimate	Std. Error	z	Pr(>|z|)	2.5 %	97.5 %
<18	0.475	0.0605	7.85	<0.001	0.356	0.593
18 to 35	0.461	0.0290	15.89	<0.001	0.404	0.518
>35	0.435	0.0338	12.84	<0.001	0.368	0.501

At first glance, it looks like the average counterfactual comparison is larger in the first age bin than in the last (0.475 vs. 0.435). It looks like the effect of `incentive` on the probability of seeking to learn one's HIV status is stronger for younger participants than for older ones.

The difference between estimated treatment effects in those two groups is

$$0.475 - 0.435 = 0.04,$$

but is this difference statistically significant? Does our data allow us to conclude that `incentive` has different effects in age subgroups?

To answer this question, we follow the process laid out in Chapter 4, and express our test as a string formula, where `b1` identifies the estimate in the first row and `b3` in the third row.

```
avg_comparisons(mod,
  hypothesis = "b1 - b3 = 0",
  variables = "incentive",
  by = "agecat")
```

Hypothesis	Estimate	Std. Error	z	Pr(>|z|)	2.5 %	97.5 %
b1-b3=0	0.04	0.0693	0.577	0.564	-0.0958	0.176

There is a numerical difference between the two estimates, but the *p* value is large. This means that we cannot reject the null hypothesis that the effect of incentive on predicted outcome is the same for participants in the two age brackets. The two estimated risk differences are not statistically distinguishable from one another.

6.6   Visualization

In many cases, data analysts will want to visualize counterfactual comparisons rather than solely report numerical summaries. Figure 6.1 already showed how one could visualize unit-level comparisons. In this section, we will see how the plot_comparisons() can be used to visualize marginal and conditional comparisons.

6.6.1   Marginal comparisons

We already know that the avg_comparisons() function can be used to compute average counterfactual comparisons, or conditional average treatment effects by subgroups of the data. For example, it allows us to calculate the average change in predicted outcome associated with a change in incentive, for different age subgroups.

```
avg_comparisons(mod, variables = "incentive", by = "agecat")
```

agecat	Estimate	Std. Error	z	Pr(>|z|)	2.5 %	97.5 %
<18	0.475	0.0605	7.85	<0.001	0.356	0.593
18 to 35	0.461	0.0290	15.89	<0.001	0.404	0.518
>35	0.435	0.0338	12.84	<0.001	0.368	0.501

Instead of displaying these results in a table, we can call the
`plot_comparisons()` function to compute the same estimates, and present
them graphically. The syntax is nearly identical.

```
library(ggplot2)
```

```
plot_comparisons(mod, variables = "incentive", by = "agecat") +
  labs(x = "Age", y = "Average risk difference")
```

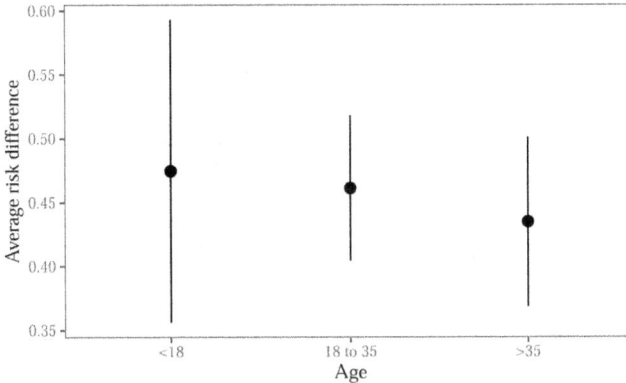

Figure 6.2: Marginal counterfactual comparisons for a change in incentive, by
age category.

This plot shows that, on average, our model expects that receiving an incentive
will have a stronger effect on the probability of seeking to learn one's HIV
status for younger individuals. Indeed, the point estimate is higher on the
left side of the plot than on the right side. However, the confidence intervals
are wide and overlapping, which suggests that the treatment effects are not
significantly different from one another.

6.6.2 Conditional comparisons

When one of the predictors of interest is continuous or when we are deal-
ing with multiple predictors, plotting conditional comparisons rather than
marginal ones can improve interpretability. The `condition` argument of the
`plot_comparisons()` function helps us construct plots where we hold some predictors at representative values, while displaying the variation in comparisons
for others.

```
library(patchwork)

p1 = plot_comparisons(mod,
  variables = "incentive",
  condition = "distance") +
  labs(y = "Conditional risk difference")

p2 = plot_comparisons(mod,
  variables = "incentive",
  condition = c("distance", "agecat")) +
  labs(y = "Conditional risk difference")

p1 + p2
```

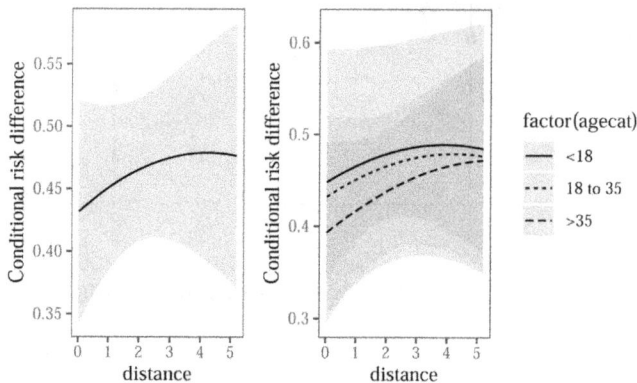

On these plots, high values on the y-axis indicate that the `incentive` has a strong effect on the predicted probability of `outcome`. The x-axis represents the distance between a participant's home and the test center, with people living closest to the test center at the left of the plots.

The first plot shows that the estimated effect of `incentive` on the predicted probability of `outcome` is smaller for individuals living closest to the test center. The second plot further indicates slight differences across age groups, with older participants reacting less strongly to the treatment. Chapter 10 shows how to explore these kinds of contrasts more formally; it shows how to design and interpret tests of treatment effect heterogeneity, using interactions and polynomials.

6.7   Summary

This chapter defined a "counterfactual comparison" as a function of two or more model-based predictions made with different predictor values. These

Table 6.2: Main arguments of the `comparisons()`, `avg_comparisons()`, and `plot_comparisons()` functions.

Argument	
`model`	Fitted model used to make counterfactual comparisons.
`variables`	Focal predictor whose association with or effect on the outcome we are interested in.
`newdata`	Grid of predictor values.
`comparison`	How the counterfactual predictions are compared: difference, ratio, lift, etc.
`vcov`	Standard errors: classical, robust, clustered, etc.
`conf_level`	Size of the confidence intervals.
`type`	Type of predictions to compare: response, link, etc.
`cross`	Estimate the effect of changing multiple variables at once (cross-contrast).
`by`	Grouping variable for average predictions.
`wts`	Weights used to compute average comparisons.
`transform`	Post-hoc transformations.
`hypothesis`	Compare different comparisons to one another, conduct linear or non-linear hypothesis tests, or specify null hypotheses.
`equivalence`	Equivalence tests
`df`	Degrees of freedom for hypothesis tests.
`numderiv`	Algorithm used to compute numerical derivatives.
`...`	Additional arguments are passed to the `predict()` method supplied by the modeling package.

comparisons measure the strength of association between variables or, if appropriate identification assumptions are met, they quantify causal effects.

The `comparisons()` function from the **marginaleffects** package computes counterfactual comparisons for a wide range of models, and `avg_comparisons()` aggregates them across units or groups. Analysts can visualize these comparisons with the `plot_comparisons()` function or standard visualization tools like **ggplot2**.

To define, compute, and interpret counterfactual comparisons, analysts must make five decisions.

First, the *Quantity*:

- Counterfactual comparisons are defined along two dimensions.
    - What change in focal predictor are we interested in? For example, do we want to estimate the effect of changing X from 0 to 1, of an increase of 1 unit or one standard deviation, or a change between two specific values. The change of focal predictor is specified by the `variables` argument.
    - What function do we use to compare counterfactual predictions? Two counterfactual predictions can be compared by a difference, ratio, lift, and many other functions. These comparisons are specified using the `comparison` argument in the `comparisons()` function.

Second, the *Predictors*:

- Counterfactual comparisons are conditional quantities. They usually depend on the values of all predictors in a model.
- Analysts can evaluate comparisons for different predictor grids, including empirical, balanced, interesting, or counterfactual grids, defined by the `newdata` argument and the `datagrid()` function.

Third, the *Aggregation*:

- Analysts can report unit-level or aggregated counterfactual comparisons:
    - Unit-level comparisons: Specific to each observation.
    - Aggregated comparisons: Average effects or associations across the population or subgroups, computed with `avg_comparisons()` and the `by` argument.

Fourth, the *Uncertainty*:

- The uncertainty around counterfactual comparisons can be estimated using classical or robust standard errors, bootstrap, or simulation-based inference.
- Standard errors and confidence intervals are handled by the `vcov` and `conf_level` arguments, or via the `inferences()` function.

Fifth, the *Test*:

- Hypothesis tests can check if counterfactual comparisons differ significantly from a null value or from one another. This is implemented with the `hypothesis` argument.
- Equivalence tests can establish whether comparisons are practically equivalent to a reference value using the `equivalence` argument.

7

Slopes

A slope measures how the predicted value of the outcome Y responds to changes in a focal predictor X, when we hold other covariates at fixed values. It is often the main quantity of interest in an empirical analysis, when a researcher wants to estimate an effect or the strength of association between two variables. Slopes belong to the same toolbox as the counterfactual comparisons explored in Chapter 6. Both quantities help us answer a counterfactual query: What would happen to a predicted outcome if one of the predictors were slightly different?

One challenge in understanding slopes is that they are formally defined in the language of calculus. This chapter does not dive deep into theory. It focuses on intuition, eschews most technical details, and provides graphical examples to make abstract ideas more tangible. Nevertheless, readers who are not comfortable with the basics of multivariable calculus may want to jump ahead to the case studies in the next part of this book.

Beyond the technical barriers, another factor that hinders our understanding of slopes is that the terminology used talk about them is woefully inconsistent. In some fields, like economics or political science, slopes are called "marginal effects," where the word "marginal" refers to the effect of an infinitesimal change in a focal predictor. In other disciplines, a slope may be called a "trend," "velocity," or "partial effect." Some analysts even use the term "slope" when they write about the raw parameters of a regression model.

In this book, we use the expressions "slope" and "marginal effect" interchangeably to mean:

> *Partial derivative of the regression equation with respect to a predictor of interest.*

Consider a simple linear model with outcome Y and predictor X.

$$Y = \beta_0 + \beta_1 X + \varepsilon \tag{7.1}$$

To find the slope of Y with respect to X, we take the partial derivative.

$$\frac{\partial Y}{\partial X} = \beta_1 \tag{7.2}$$

DOI: 10.1201/9781003560333-7

The slope of Equation 7.1 with respect to X is thus equivalent to β_1, the only coefficient in the equation. This identity explains why some analysts refer to regression coefficients as slopes. It also gives us a clue for interpretation.

In simple linear models, a one-unit increase in X is associated with a β_1 change in Y, holding other modeled covariates constant. Likewise, we will often be able to interpret the slope, $\frac{\partial Y}{\partial X}$, as the effect of a one-unit change in X on the predicted value of Y. This interpretation is useful as a first cut, but it comes with caveats.

First, given that slopes are defined as derivatives, that is, in terms of an infinitesimal change in X, they can only be constructed for continuous numeric predictors. Analysts who are interested in the effect of a change in a categorical predictor should turn to counterfactual comparisons.[1]

Second, the partial derivative measures the slope of Y's tangent at a specific point in the predictor space. Thus, the interpretation of a slope as the effect of a one-unit change in X must be understood as an approximation, valid in a small neighborhood of the predictors, for a small change in X. This approximation may not be particularly good when the regression function is non-linear, or when the scale of X is small.

Finally, the foregoing discussion implies that slopes are conditional quantities, in the sense that they typically depend on the value of X, but also on the values of all the other predictors in a model. Every row of a dataset or grid has its own slope. In that respect, slopes are like the predictions and counterfactual comparisons from chapters 5 and 6.

The rest of this chapter proceeds as before, by answering the five questions of our conceptual framework: (1) quantity, (2) predictors, (3) aggregation, (4) uncertainty, and (5) test.

7.1   Quantity

A slope characterizes the strength and the direction of association between a predictor and an outcome, holding other covariates constant. To get an intuitive feel for what this means in practice, it is useful to consider a few graphical examples. In this section, we will see that the slope of a prediction curve tells us how it behaves as we move from left to right along the x-axis, that is, it tells us whether the curve is increasing, decreasing, or flat. This information is useful to understand the relationships between variables of interest in a regression model.

---

[1]Note that the `slopes()` function from the `marginaleffects` package automatically reverts to counterfactual comparisons if some of the predictors are categorical.

7.1.1 Slopes of simple functions

The top row of Figure 7.1 shows three functions of X. The bottom row traces the derivatives of each of those functions. By examining the values of derivatives at any given point, we can determine precisely where the corresponding functions are rising, falling, or flat.

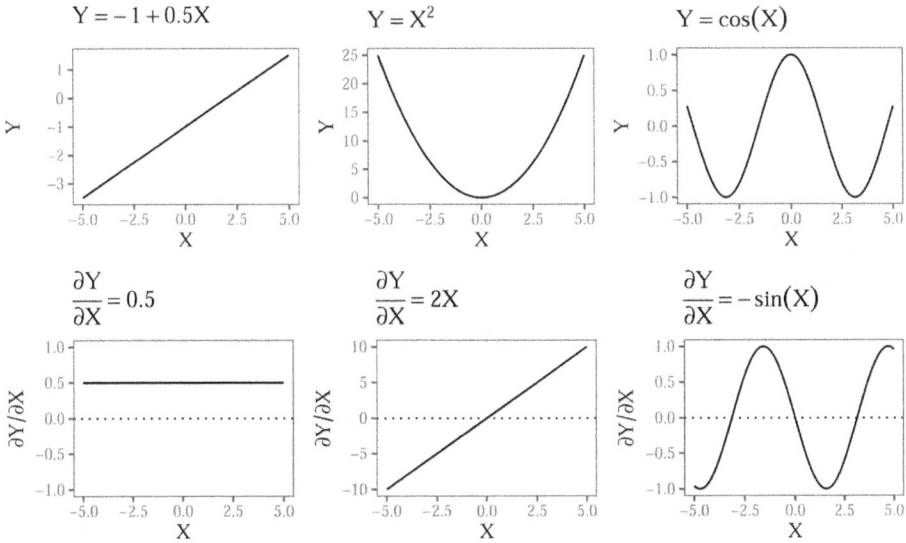

Figure 7.1: Three functions and their derivatives.

First, consider the top left panel of Figure 7.1, which displays the function $Y = -1 + 0.5X$, a straight line with a positive slope. As we move from left to right, the line steadily rises. The effect of X is constant across the full range of X values. For every one-unit increase in X, Y increases by 0.5 units. Since the slope does not change, the derivative of this function is constant (bottom left panel): $\frac{\partial Y}{\partial X} = 0.5$. This means that the association between X and Y is always positive, and that the strength of association between X and Y is of constant magnitude, regardless of the baseline value of X.

Now, consider the top middle panel of Figure 7.1, which displays the function $Y = X^2$, a symmetric parabola that opens upward. As we move from left to right, the function initially decreases, until it reaches its minimum at $X = 0$. In the left part of the plot, the relationship between X and Y is negative: an increase in X is associated to a decrease in Y. When X reaches 0, the curve becomes flat: the slope of its tangent line is zero. This means that an infinitesimal change in X would have (essentially) no effect on Y. As X continues to increase beyond zero, the function starts increasing, with a positive slope. The derivative of this function captures this behavior. When $X < 0$, the derivative is $\frac{\partial Y}{\partial X} = 2X < 0$, which indicates that the relationship between X and Y is negative. When $X = 0$, the derivative is $2X = 0$, which means that the association between X and Y is null. When the $X > 0$, the derivative is positive, which implies that an increase in X is associated with an increase in Y. In short, the sign and strength of the relationship between X and Y are heterogeneous; they depend on the baseline value of X. This will be an important insight to keep in mind for the case study in Chapter 10, where we explore the use of interactions and polynomials to study heterogeneity and non-linearity.

The third function in Figure 7.1, $Y = \cos(X)$, oscillates between positive and negative values, forming a wave-like curve. Focusing on the first section of this curve, we see that $\cos(X)$ decreases as we move from left to right, reaches a minimum at $-\pi$, and then increases until $X = 0$. The derivative plotted in the bottom right panel captures this trajectory well. Indeed, whenever $-\sin(X)$ is negative, we see that $\cos(X)$ points downward. When $-\sin(X) = 0$, the $\cos(X)$ curve is flat and reaches a minimum or a maximum. When $-\sin(X) > 0$, the $\cos(X)$ function points upward.

These three examples show that the derivative of a function precisely characterizes both the strength and the direction of association between a predictor and an outcome. It tells us, at any given point in the predictor space, if increasing X should result in a decrease or an increase in Y.

7.1.2   Slope of a logistic function

Let us now consider a more realistic example, similar to the curves we are likely to fit in an applied regression context.

$$Pr(Y = 1) = g\left(\beta_1 + \beta_2 X\right), \tag{7.3}$$

where g is the logistic function, β_1 is an intercept, β_2 a regression coefficient, and X is a numeric predictor. Imagine that the true parameters are $\beta_1 = -1$ and $\beta_2 = 0.5$. We can simulate a dataset with a million observations that follow this data generating process.

```
library(marginaleffects)
library(patchwork)
set.seed(48103)
N = 1e6
X = rnorm(N, sd = 2)
p = plogis(-1 + 0.5 * X)
Y = rbinom(N, 1, p)
dat = data.frame(Y, X)
```

Now, we fit a logistic regression model with the `glm()` function and print the coefficient estimates.

```
mod = glm(Y ~ X, data = dat, family = binomial)
b = coef(mod)
b
```

```
(Intercept)          X
 -0.9981282    0.4993960
```

The estimated coefficients are very close to the true values we encoded in the data-generating process.

To visualize the outcome function and its derivative, we use the `plot_predictions()` and `plot_slopes()` functions. The latter is very similar to the `plot_comparisons()` function introduced in Section 6.6. The `variables` argument identifies the focal predictor with respect to which we are computing the slope, and the `condition` argument specifies which variable should be displayed on the x-axis. To display plots on top of one another, we use the `/` operator supplied by the `patchwork` package.

```
p_function = plot_predictions(mod, condition = "X")
p_derivative = plot_slopes(mod, variables = "X", condition = "X")
p_function / p_derivative
```

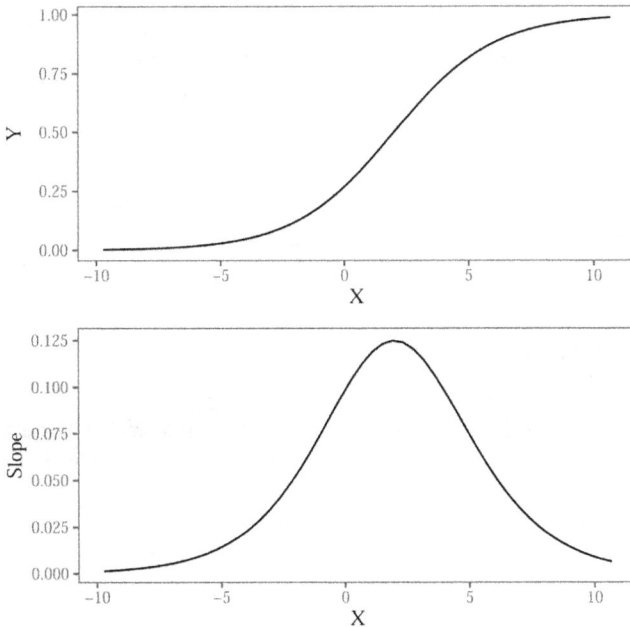

Figure 7.2: Logistic function and its derivative.

The key insight from Figure 7.2 is that the slope of the logistic function is far from constant. Increasing X, moving from left to right on the graph, has a different effect on the estimated probability that $Y = 1$, depending on our position on the horizontal axis. When the initial value of X is small (left side of the figure), an increase in X results in a small change in $Pr(Y = 1)$: the prediction curve is relatively flat and the derivative is close to zero. At intermediate values of X (middle of the figure), a change in X results in a large increase in $Pr(Y = 1)$: the outcome prediction curve is steep, and its derivative is positive and large. When the initial value of X is large (right side of the figure), a change in X does not change $Pr(Y = 1)$ much: the outcome curve is flat and the derivative small again. In sum, the slope characterizes the strength of association between X and Y for different baseline values of the predictors.

To estimate the slope, we can proceed analytically, using the chain rule of differentiation. Taking the derivative of Equation 7.3 with respect to X gives us

$$\frac{\partial Pr(Y = 1)}{\partial X} = \beta_2 \cdot g'(\beta_1 + \beta_2 X), \tag{7.4}$$

where g' is the logistic density function or the derivative of g.[2] To obtain an estimate of the slope, we simply plug-in our coefficient estimates into Equation 7.4. Importantly, whenever we evaluate a slope, we must explicitly state the baseline values of the predictors of interest. The slope of Y with respect to X will be different when X takes on different values. For example, the estimated $\frac{\partial Pr(Y=1)}{\partial X}$ for $X \in \{-5, 0, 10\}$ are

```
data.frame(
  "dY/dX|X=-5" = b[2] * dlogis(b[1] + b[2] * -5),
  "dY/dX|X=0" = b[2] * dlogis(b[1] + b[2] * 0),
  "dY/dX|X=10" = b[2] * dlogis(b[1] + b[2] * 10)
)
```

```
dY/dX|X=-5  dY/dX|X=0  dY/dX|X=10
0.0142749  0.09827211  0.008856198
```

Alternatively, we could compute the same estimates with the `slopes()` function from `marginaleffects`. `slopes()` is very similar to the `comparisons()` function introduced in Chapter 6. The `variables` argument specifies the focal predictor, and `newdata` specifies the values of the other predictors at which we want to evaluate the slope.

```
slopes(mod,
  variables = "X",
  newdata = datagrid(X = c(-5, 0, 10)))
```

---

[2]The logistic density is a relatively complex function, but thankfully it is implemented in R via the `dlogis()` function.

X	Estimate	Std. Error	z	Pr(>|z|)	2.5 %	97.5 %
-5	0.01427	7.24e-05	197.0	<0.001	0.01413	0.01442
0	0.09827	2.61e-04	376.6	<0.001	0.09776	0.09878
10	0.00886	8.98e-05	98.6	<0.001	0.00868	0.00903

The numeric estimates obtained for this contrived example are extremely similar to those we computed analytically, which is expected. In the rest of this chapter, we will explore how to use `slopes()` and its siblings `avg_slopes()` and `plot_slopes()`, to estimate and visualize a variety of slope-related quantities of interest.

For illustration, we will look at a variation on a model considered in previous chapters, using data from Thornton (2008). As before, the dependent variable is a dummy variable called `outcome`, equal to 1 if a study participant sought to learn their HIV status at a test center, and 0 otherwise. Since we are studying partial derivatives, the focal predictor must be a continuous numeric variable. Here, we choose `distance`, a measure the distance between the participant's residence and the test center. Our model also includes `incentive`, a binary variable which we treat as a control.

The model we specify includes multiplicative interactions between `incentive`, `distance`, and its square. These interactions give the model a bit more flexibility to detect non-linear patterns in the data, but their real purpose is pedagogical: they show that the post-estimation workflow introduced in previous chapters can be applied in straightforward fashion to more complex models. Once again, we will ignore individual parameter estimates, and focus on quantities of interest with more intuitive substantive meanings.

```
dat = get_dataset("thornton")
mod = glm(outcome ~ incentive * distance * I(distance^2),
  data = dat, family = binomial)
summary(mod)
```

```
Coefficients:
                                  Estimate Std. Error z value Pr(>|z|)
(Intercept)                        0.42793    0.38264   1.118   0.2634
incentive                          1.79455    0.47386   3.787   <0.001
distance                          -1.48974    0.63658  -2.340   0.0193
I(distance^2)                      0.56304    0.29767   1.891   0.0586
incentive:distance                 0.49785    0.76074   0.654   0.5128
incentive:I(distance^2)           -0.25193    0.34657  -0.727   0.4673
distance:I(distance^2)            -0.06694    0.03981  -1.682   0.0926
incentive:distance:I(distance^2)   0.03426    0.04544   0.754   0.4509
```

7.2 Predictors

Slopes can help us answer questions such as:

How does the predicted outcome $\hat{Y}$ change when the focal variable X increases by a small amount and the adjustment variables $Z_1, Z_2, \ldots, Z_n$ are held at specific values?

Answering questions like this requires us to select both the focal variable and the values of the adjustment variables where we want to evaluate the slope.

7.2.1 Focal variable

The focal variable is the predictor of interest. It is the variable whose association with (or effect on) Y we wish to estimate. The focal variable is the predictor in the denominator of the partial derivative notation $\frac{\partial Y}{\partial X}$.

The `slopes()` function accepts a `variables` argument, which serves to specify a focal predictor. For example, the following code estimates the partial derivative of the outcome equation encoded in the `mod` object, with respect to the `distance` focal predictor.

```
slopes(mod, variables = "distance")
```

7.2.2 Adjustment variables

Slopes are conditional quantities, in the sense that they typically depend on the values of all the variables on the right-hand side of a regression equation. Thus, every predictor profile—or combination of predictor values—will be associated with its own slope. Every row in a grid will have its own slope. Section 3.2 described different grids of predictor values, that is, different combinations of unit characteristics that could be of interest.

One common type of grid is the "interesting" or "user-specified" grid, which collects combinations of predictor values that hold particular scientific or domain-specific interest. For example, imagine that we are specifically interested in measuring the association between `distance` and `outcome` for an individual who is part of the treatment group (`incentive` is 1) and who lives at a distance of 1 from the test center. We can use the `datagrid()` helper function to specify a grid of interesting predictor values, and then pass that grid to the `newdata` argument of the `slopes()` function. The function will then return a "slope at user-specified values" or "marginal effect at interesting values."

```
library(marginaleffects)
slopes(mod,
  variables = "distance",
  newdata = datagrid(incentive = 1, distance = 1))
```

incentive	distance	Estimate	Std. Error	z	Pr(>|z|)	2.5 %	97.5 %
1	1	−0.0694	0.021	−3.3	<0.001	−0.111	−0.0282

Our model suggests that for an individual with baseline characteristics `incentive=1` and `distance=1`, a one-unit increase in `distance` is associated with a decrease of about 6.9 percentage points in the probability that a study participant will seek to learn their HIV status. This interpretation is authorized by the fact that the printed estimate corresponds to the slope of the tangent of the outcome curve at the specified point in the predictor space. But it is important to remember that this interpretation is a linear approximation, only valid for small changes in the focal predictor.

Instead of manually specifying the baseline values of all predictors, we can set `newdata="mean"` to compute a "marginal effect at the mean" or a "marginal effect at representative values." This is the slope of the regression equation with respect to the focal predictor, for an individual whose characteristics are exactly average (or modal) on all predictors.

```
slopes(mod, variables = "distance", newdata = "mean")
```

Estimate	Std. Error	z	Pr(>|z|)	2.5 %	97.5 %
−0.024	0.0135	−1.78	0.0758	−0.0505	0.0025

As noted in previous chapters, computing estimates at the mean is computationally efficient, but it may not be particularly informative when the perfectly average individual is not realistic or substantively interesting. The choice of reporting a slope at the mean may seem innocuous, but it is not. In fact, our slope at the mean is quite different from the slope that we computed above for specified values of the covariates: −0.024 vs. −0.069, or about a third of the size. This emphasizes the crucial importance of grid definition.

Finally, instead of computing slopes for just a few profiles, we can get them for every observation in the original sample. These "unit-level marginal effects" or "partial effects" are the default output of the `slopes()` function.

```
slopes(mod, variables = "distance")
```

Estimate	Std. Error	z	Pr($>$|z|)	2.5 %	97.5 %
−0.23004	0.0915	−2.5141	0.01193	−0.4094	−0.0507
−0.16236	0.0598	−2.7140	0.00665	−0.2796	−0.0451
0.00663	0.0291	0.2280	0.81962	−0.0503	0.0636
0.00895	0.0326	0.2743	0.78389	−0.0550	0.0729
−0.06896	0.0278	−2.4821	0.01306	−0.1234	−0.0145
		2815 rows omitted			
−0.05763	0.0165	−3.4881	< 0.001	−0.0900	−0.0252
−0.07078	0.0216	−3.2793	0.00104	−0.1131	−0.0285
−0.05771	0.0166	−3.4864	< 0.001	−0.0902	−0.0253
−0.04587	0.0132	−3.4627	< 0.001	−0.0718	−0.0199
−0.00147	0.0160	−0.0921	0.92662	−0.0328	0.0299

7.3   Aggregation

A dataset with one marginal effect estimate per unit of observation is a bit unwieldy and difficult to interpret. Instead of presenting a large set of estimates, many analysts prefer to report the "average marginal effect" (or "average slope"), that is, the average of all the unit-level estimates. The average slope can be obtained in two steps, by computing unit-level estimates and then their average. However, it is more convenient to call the `avg_slopes()` function.

```
avg_slopes(mod, variables = "distance")
```

Estimate	Std. Error	z	Pr($>$|z|)	2.5 %	97.5 %
−0.0516	0.00919	−5.61	<0.001	−0.0696	−0.0335

Note that there is a nuanced distinction to be drawn between the "marginal effect at the mean" shown in the previous section, and the "average marginal effect" presented here. The former is calculated based on a single individual whose characteristics are exactly average or modal in every dimension. The latter is calculated by taking the average of slopes for all the observed data points in the dataset used to fit a model. These two options will not always be equivalent; they can yield numerically and substantively different results. In general, the marginal effect at the mean might be useful when there are computational constraints. The average marginal effect will be useful when the dataset adequately represents the distribution of predictors in the population, in which case marginalizing across the distribution can give us a good one-number summary.

Sometimes, the average slope ignores important patterns in the data. For instance, we can imagine that `distance` has a different effect on `outcome` for individuals who receive a monetary `incentive` and for those who do not. To explore this kind of heterogeneity in the association between independent and dependent variables, we can use the `by` argument. This will compute "conditional average marginal effects," or average slopes by subgroup.

The following results, for instance, show the average slope (strength of association) between `distance` and the probability that a study participant will travel to the clinic to learn their test result.

```
avg_slopes(mod, variables = "distance", by = "incentive")
```

incentive	Estimate	Std. Error	z	Pr(>|z|)	2.5 %	97.5 %
0	−0.0818	0.02464	−3.32	<0.001	−0.1301	−0.0335
1	−0.0430	0.00952	−4.52	<0.001	−0.0617	−0.0244

Indeed, it seems that distance discourages travel to the clinic for individuals in both groups, but that the strength of the negative association between `distance` and `outcome` is stronger when `incentive=0`. In Section 7.5, we will formally test if the difference between these estimates is statistically significant.

7.4   Uncertainty

As with all other quantities estimated thus far, we can compute estimates of uncertainty around slopes using various strategies: Classical or robust standard errors, bootstrapping, simulation-based inference, etc. To do this, we use the `vcov` and `conf_level` arguments, or the `inferences()` function. Here are two simple examples: the first command clusters standard errors by village, while the second applies a non-parametric bootstrap.

```
avg_slopes(mod, variables = "distance", vcov = ~village)
```

Estimate	Std. Error	z	Pr(>|z|)	2.5 %	97.5 %
−0.0516	0.0112	−4.61	<0.001	−0.0734	−0.0297

```
avg_slopes(mod, variables = "distance") |> inferences(method = "boot")
```

Estimate	2.5 %	97.5 %
−0.0516	−0.0676	−0.0339

The estimates are the same, but the confidence intervals differ, depending on the uncertainty quantification strategy.

7.5   Test

Chapter 4 explained how to conduct complex null hypothesis tests on any quantity estimated by the marginaleffects package. The approach introduced in that chapter is applicable to slopes, as it was to model coefficients, predictions, and counterfactual comparisons.

Recall the example from Section 7.3, where we computed the average slopes of outcome with respect to distance, for each incentive subgroup.

```
avg_slopes(mod,
  variables = "distance",
  by = "incentive")
```

incentive	Estimate	Std. Error	z	Pr($>$|z|)	2.5 %	97.5 %
0	−0.0818	0.02464	−3.32	<0.001	−0.1301	−0.0335
1	−0.0430	0.00952	−4.52	<0.001	−0.0617	−0.0244

At first glance, these two estimates appear different. It seems that distance is more discouraging to people who do not receive an incentive. One might thus infer that money can mitigate the barrier posed by distance.

To ensure that this observation is not simply the product of sampling variation, we must formally check if the two estimates in the table are statistically distinguishable. To do this, we execute the same command as before, but add a hypothesis argument with an equation-like string. That equation specifies our null hypothesis: there is no difference between the first estimate (b1) and the second (b2).

```
avg_slopes(mod,
  variables = "distance",
  by = "incentive",
  hypothesis = "b1 - b2 = 0")
```

Hypothesis	Estimate	Std. Error	z	Pr(>|z|)	2.5 %	97.5 %
b1-b2=0	−0.0388	0.0264	−1.47	0.142	−0.0905	0.013

The difference between our two estimates is about −0.039, which suggests that the slope is flatter in the group with incentive. In other words, the association between `distance` and `outcome` is weaker when `incentive=1` than when `incentive=0`. However, this difference is not statistically significant ($p=0.142$). Therefore, we cannot reject the null hypothesis of homogeneity in the effect of distance on the predicted probability that participants will seek their test results.

7.6   Visualization

When interpreting the results obtained by fitting a model, it is important to keep in mind the distinction between predictions and slopes. Predictions give us the level of an expected outcome, for given values of the predictors. Slopes capture how the expected outcome changes in response to a change in a focal variable, while holding adjustment variables constant. Figure 7.3 illustrates these concepts in two panels.

The top panel shows model-based predictions for the `outcome` variable at different values of `distance`, while holding other predictors at their mean or mode. As we move from left to right along the x-axis, as `distance` from the test center increases, the predicted probability of `outcome=1` declines, flattens, and then declines again.

This pattern is mirrored in the bottom panel of Figure 7.3, which shows the slope of predicted `outcome` with respect to `distance`, holding adjustment variables at their mean or mode. For most values of `distance`, the slope is negative, which means that `distance` has a negative effect on the predicted probability that `outcome` equals 1. When `distance` is around 3 or 4, the association between `distance` and `outcome` is flat or null. At that point, small increases in `distance` no longer have much of an effect on predicted `outcome`. When the slope is below zero (dotted line), the prediction curve in the top panel goes down. When the slope is at zero, the prediction curve is flat.

To draw the two plots in Figure 7.3, we call the `plot_predictions()` and `plot_slopes()` functions, and combine the outputs in a single figure using the / operator from the `patchwork` package.

```
library(ggplot2)
library(patchwork)

p1 = plot_predictions(mod, condition = "distance") +
    labs(y = "Predicted Pr(Outcome=1)")

p2 = plot_slopes(mod, variables = "distance", condition = "distance") +
    geom_hline(yintercept = 0, linetype = "dotted") +
    labs(y = "dY/dX")

p1 / p2
```

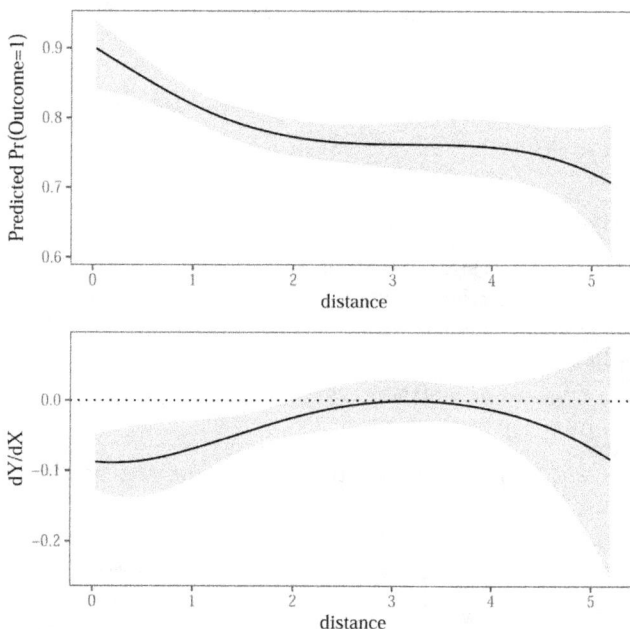

Figure 7.3: Model-based predictions (top) and slopes (bottom).

`plot_slopes()` can display several slopes based on multiple conditions. For example, we can show how the association between `distance` and `outcome` varies based on `incentive` and `distance`.

```
plot_slopes(mod,
  variables = "distance",
  condition = c("distance", "incentive"))
```

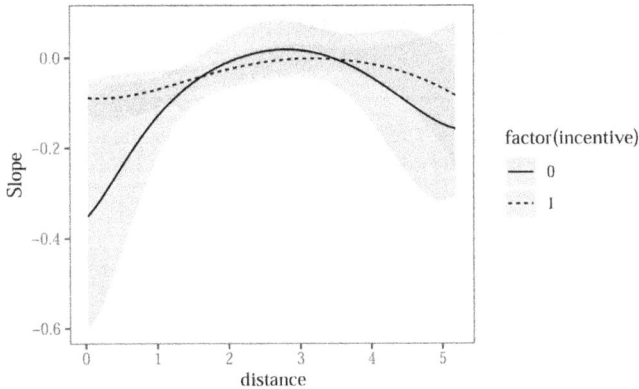

We can also compute *marginal* slopes, that is, the average of individual-level slopes by subgroup. To do this, we use the `by` argument. The next plot shows the same results we explored in Section 7.5. The negative association between `distance` and `outcome` seems weaker when `incentive=1` than when `incentive=0`, but the confidence intervals are wide, so we cannot reject the null hypothesis that the two average slopes are equal.

```
plot_slopes(mod,
  variables = "distance",
  by = "incentive")
```

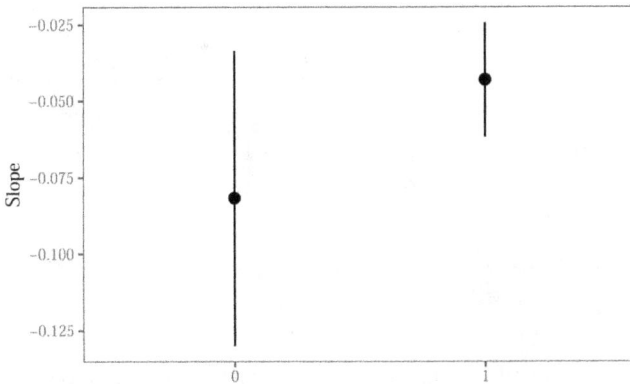

7.7 Summary

This chapter defined a "slope" as the partial derivative of the regression equation with respect to a predictor of interest. It is a measure of association between two variables, or of the effect of one variable on another, holding other predictors constant.

Table 7.1: Main arguments of the `slopes()`, `avg_slopes()`, and `plot_slopes()` functions.

Argument	
`model`	Fitted model used to make counterfactual slopes.
`variables`	Focal predictor whose association with or effect on the outcome we are interested in.
`newdata`	Grid of predictor values.
`slope`	Choice between derivative or (semi-)elasticity.
`vcov`	Standard errors: Classical, robust, clustered, etc.
`conf_level`	Size of the confidence intervals.
`type`	Type of predictions to compare: response, link, etc.
`by`	Grouping variable for average predictions.
`wts`	Weights used to compute average slopes.
`hypothesis`	Compare different slopes to one another, conduct linear or non-linear hypothesis tests, or specify null hypotheses.
`equivalence`	Equivalence tests
`df`	Degrees of freedom for hypothesis tests.
`numderiv`	Algorithm used to compute numerical derivatives.
`...`	Additional arguments are passed to the `predict()` method supplied by the modeling package.

The `slopes()` function from the **marginaleffects** package computes slopes for a wide range of models. `avg_slopes()` aggregates slopes across units or groups. `plot_slopes()` displays slopes visually.

To clearly define slopes and attendant tests, analysts must make five decisions.

First, the *Quantity*.

- A slope is always computed with respect to a focal predictor, whose effect on (or association with) the outcome we wish to estimate. In **marginaleffects** functions, the focal variable is specified using the `variables` argument.
- A slope can roughly be interpreted as the effect of a one-unit change in the focal predictor on the predicted outcome. However, this interpretation is a linear approximation, valid only in a small neighborhood of the predictors.

Second, the *Predictors*.

- Slopes are conditional quantities, meaning that they will typically vary based on the values of all predictors in a model. Every row of a dataset has its own slope.
- Analysts can compute slopes for different combinations of predictor values—or grids: empirical, interesting, representative, balanced, or counterfactual.
- The predictor grid is defined by the `newdata` argument and the `datagrid()` function.

Third, the *Aggregation*.

- To simplify the presentation of results, analysts can report average slopes. Different aggregation schemes are available:
  - Unit-level slopes (no aggregation)
  - Average slopes
  - Average slopes by subgroup
  - Weighted average of slopes
- Slopes can be aggregated using the `avg_slopes()` function and the `by` argument.

Fourth, the *Uncertainty*.

- In `marginaleffects`, the `vcov` argument allows analysts to report classical, robust, or clustered standard errors for slopes.
- The `inferences()` function can compute uncertainty intervals via bootstrapping or simulation-based inference.

Fifth, the *Test*.

- A null hypothesis test evaluates whether a slope (or a function of slopes) is significantly different from a null value. For example, we may use a null hypothesis test to check if treatment effects are equal in subgroups of the sample. Null hypothesis tests are conducted using the `hypothesis` argument.
- An equivalence test evaluates whether a slope (or a function of slopes) is similar to a reference value. Equivalence tests are conducted using the `equivalence` argument.

Part III

Case studies

8

Causal inference with G-computation

Randomized experiments are often considered the gold standard research design for causal inference. Assigning individuals to different treatments at random ensures that, on average and under ideal conditions, the groups we compare have similar background characteristics. This overall similarity gives us more confidence that any differences to emerge between groups are caused by the treatments of interest. Randomized experiments are powerful tools but, unfortunately, they are often impractical or unethical. In such cases, we must turn to observational data.

Drawing causal inference from observational data can be difficult. Often, factors beyond the analyst's control influence both the exposure to treatment and the outcome of interest. Such confounding variables can introduce bias in our estimates of the treatment effect.

Imagine a study designed to assess the effect of daily vitamin D supplementation on bone density. If an analyst finds that individuals taking vitamin D have higher bone density, they may be tempted to conclude that supplementation directly enhances bone health. However, people who choose to take vitamin D might also be more health-conscious and engage in other bone-strengthening activities, such as exercising or eating calcium-rich foods. These confounding factors can obscure the true causal effect. They can introduce bias in the analyst's estimates.

G-computation[1] is a method developed to draw causal inference from observational data. The idea is simple. First, we fit a statistical model which controls for confounders. Then, we use the fitted model to "predict" or "impute" what would happen to an individual under alternative treatment scenarios. Finally, we compare counterfactual predictions to derive an estimate of the treatment effect (Hernán and Robins, 2020).

As we will see below, G-computation estimates are equivalent to the counterfactual predictions and comparisons discussed in Chapters 5 and 6. They are also

---

[1]The estimation procedure described in this chapter is sometimes referred to as the "Parametric G-Formula."

closely related to the estimates we can obtain by Inverse Probability Weighting (IPW), another popular causal inference tool. IPW and G-computation impose similar identification assumptions, with one key difference: the former models the process that determines who gets treated, whereas the latter models the outcome variable.[2]

The next section introduces some key estimands that analysts can target via G-computation: the average treatment effect (ATE), average treatment effect on the treated (ATT), and average treatment effect on the untreated (ATU). Subsequent sections present the estimation procedure as a sequence of three steps: model, impute, and compare.

8.1   Treatment effects: ATE, ATT, ATU

Our goal is to estimate the effect of a treatment D on an outcome Y. To express this effect as a formal estimand, and to understand the assumptions required to target that estimand, we need to introduce a bit of notation from the potential outcomes tradition of causal inference.[3]

The first term we need to define represents the value of the treatment, or explanatory variable. When $D_i = 1$, the individual i is assigned to the treatment group. When $D_i = 0$, i is part of the control group.

The second term represents the outcome variable. Here, it is essential to draw a distinction between the *actual* value of the outcome, and its *potential* values. The actual outcome is what we observe for individual i in our dataset: Y_i. In contrast, the potential outcomes are written with superscripts: Y_i^0 and Y_i^1. They represent the outcomes that would occur in counterfactual worlds where i receives different treatments. Y_i^0 is the value of the outcome in a hypothetical world where i belongs to the control group. Y_i^1 is the value of the outcome in a hypothetical world where i belongs to the treatment group. Crucially, only one potential outcome can ever be observed at any given time, for any given individual. Since i cannot be part of both the treatment and control groups simultaneously, we can never measure the values of both Y_i^1 and Y_i^0. We can only measure the potential outcome that corresponds to the treatment actually assigned to individual i.

---

[2]G-computation and IPW can be combined to create a "doubly-robust" estimator with useful properties. A comprehensive presentation of IPW and doubly-robust estimation lies outside the scope of this book. Interested readers can refer to the marginaleffects.com website for tutorials, and to Chatton and Rohrer (2024) for intuition and comparisons.

[3]For a brief discussion, see Section 2.1.3. For more detailed presentations, see Imbens and Rubin (2015), Morgan and Winship (2015), Ding (2024).

The ambitious analyst would love to estimate the *Individual Treatment Effect* (ITE), defined as the difference between outcomes for i under different treatment regimes: $Y_i^1 - Y_i^0$. Unfortunately, since we can only observe one of the two potential outcomes, the ITE is impossible to compute. This is what some have called the *fundamental problem of causal inference*.

To circumvent this problem, we turn away from individual-level estimands, and consider three aggregated quantities of interest: ATE, ATT, and ATU. These three quantities are related, but they invite different interpretations and impose different assumptions (Greifer and Stuart, 2021).

8.1.1 Interpretation

Our three estimands are defined as follows:

$$E[Y_i^1 - Y_i^0] \tag{ATE}$$
$$E[Y_i^1 - Y_i^0 | D_i = 1] \tag{ATT}$$
$$E[Y_i^1 - Y_i^0 | D_i = 0] \tag{ATU}$$

The ATE estimates the average effect of the treatment on the outcome in the entire study population. The ATE helps us answer this question: Should a program or treatment be implemented universally?

The ATT estimates the average effect of the treatment on the outcome in the subset of study participants who actually received it. The ATT helps us answer this question: Should a program or treatment be withheld from the people currently receiving it?

The ATU estimates the average effect of the treatment on the subgroup of participants who did not, in fact, get it. The ATU helps us answer this question: Should a program or treatment be extended to those who did not initially receive it?

8.1.2 Assumptions

Estimating the ATE, ATT, or ATU requires us to accept four assumptions, with varying levels of stringency (Greifer and Stuart, 2021; Chatton and Rohrer, 2024).

First, *conditional exchangeability* or no unmeasured confounding. This assumption states that potential outcomes must be independent of treatment assignment, conditional on a set of control variables (Z_i). More formally, we can write $Y_i^1, Y_i^0 \perp D_i \mid Z_i$.

Imagine a person who considers entering a training program designed to improve their likelihood of finding a job. If that person knows they will be employed regardless of whether they receive training, then they are unlikely to enroll in the program: if $Y_i^1 = Y_i^0$, then $D_i = 0$. Similarly, imagine a doctor who recommends a certain medication only to the patients who are most likely to benefit from it: if $Y_i^1 > Y_i^0$, then $D_i = 1$. In both cases, potential outcomes are linked to the treatment assignment mechanism: $D_i \not\perp Y_i^1, Y_i^0$. In both cases, the conditional exchangeability assumption is violated, unless the analyst can control for all relevant confounders.

When estimating the ATE, conditional exchangeability must hold across the entire population. When estimating the ATT, however, this stringent assumption can be relaxed somewhat. The intuition is straightforward: since the ATT is computed solely based on units in the treatment group, we can observe all instances of the Y_i^1 potential outcomes. Thus, to estimate the ATT, we only need to make assumptions about the unobserved Y_i^0 potential outcomes; we only need to adjust for confounders that link treatment assignment to participants' outcomes under the control condition. Analogously, estimating the ATU only requires conditional exchangeability to hold in the treatment condition.

Second, *positivity*. This assumption requires study participants to have a non-zero probability of belonging to each treatment arm: $0 < P(D_i = 1 \mid Z_i) < 1$. In practice, this condition can be violated for a number of reasons, such as when certain individuals fail to meet eligibility criteria to receive a certain treatment, but are still part of the control group.

When we estimate the ATE, positivity must hold across the entire sample. However, when we estimate the ATT, the positivity assumption can be relaxed somewhat: $P(D_i = 1|Z_i) < 1$. This new condition means that no participant can be certain to receive the treatment, but that some participants can be ruled ineligible. Analogously, we can estimate the ATU by accepting a weaker form of positivity, where every participant must have some chance to be treated: $P(D_i = 1|Z_i) > 0$.

Third, *consistency*.[4] This assumption requires the intervention to be well-defined; there must be no ambiguity about what constitutes "treatment" and "control." For example, in a medical study, consistency may require that all participants in the treatment group be administered the same drug, under the same conditions, with the same dosage. In the context of a job training intervention, consistency might require strict adherence to a specific program design, with fixed hours of in-person teaching, number of assignments, internship requirements, etc.

---

[4]Formally, if $D_i = 1$ then $Y_i = Y_i^1$, and if $D_i = 0$ then $Y_i = Y_i^0$. The consistency and non-interference assumptions are often stated together under the label "Stable Unit Treatment Value Assumption" or SUTVA.

Fourth, *non-interference*. This final assumption says that potential outcomes for one participant should not be affected by the treatment assignments of other individuals. Assigning one person (or group) to a treatment regime cannot lead to contagion or generate externalities for other participants in the study.

When these four assumptions hold, we can estimate the effect of a treatment via G-computation. The remainder of this section describes this estimation strategy as a series of three steps: model, impute, and compare.

8.1.3 Model

To illustrate how to estimate the ATE, ATT, and ATU using G-computation, let us consider this study by Imbens et al. (2001): *Estimating the Effect of Unearned Income on Labor Earnings, Savings, and Consumption: Evidence from a Survey of Lottery Players*. In that article, the authors conducted a survey to estimate the effect of winning large lottery prizes on economic behavior. They studied a range of outcomes for different subgroups of the population. This chapter focuses on the effect of winning a big lottery prize on labor earnings.

The treatment is a binary variable equal to one if an individual won a big lottery prize, and zero otherwise (D). The outcome is the average labor earnings over the subsequent six years (Y). Winning lottery numbers are chosen randomly, but the probability that any given person wins a prize is not purely random: it depends on how many tickets that person bought (L). Finally, we know that socio-demographic characteristics (Z), like age, education, and employment status, can influence both the number of tickets that a person buys and their labor earnings. These causal relationships are illustrated in the graph below.

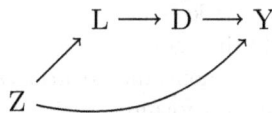

$$L \longrightarrow D \longrightarrow Y$$

$$Z$$

Imbens et al. (2001) collected survey responses to measure the variables in this graph. Their dataset includes a variety of information about 437 respondents, including 194 who won small lottery prizes, and 43 who won big prizes. For simplicity, we will only compare individuals who won a big prize to those who won nothing at all.

```
library(marginaleffects)
dat = get_dataset("lottery")
dat = subset(dat, win_big == 1 | win == 0)

head(dat, n = 2)
```

```
# A data frame: 2 x 26
    age college education    man prize tickets    win win_big win_small   work
*  <dbl>   <dbl>       <dbl> <dbl> <dbl>   <dbl> <dbl>   <dbl>     <dbl> <dbl>
1     76       0          12     0     0       2     0       0         0     0
2     59       1          16     1     0       2     0       0         0     1
# i 16 more variables: year <dbl>, earnings_pre_avg <dbl>,
#    earnings_post_avg <dbl>, earnings_pre_1 <dbl>, earnings_pre_2 <dbl>,
#    earnings_pre_3 <dbl>, earnings_pre_4 <dbl>, earnings_pre_5 <dbl>,
#    earnings_pre_6 <dbl>, earnings_post_1 <dbl>, earnings_post_2 <dbl>,
#    earnings_post_3 <dbl>, earnings_post_4 <dbl>, earnings_post_5 <dbl>,
#    earnings_post_6 <dbl>, earnings_post_7 <dbl>
```

The first G-computation step is to define an outcome equation, that is, to specify a parametric regression model for the dependent variable. The primary goal of this regression model is to control for confounders, that is, to ensure that we satisfy the conditional exchangeability assumption.

As noted above, the process that determines which lottery tickets are winning is purely random, but the probability that any given individual wins depends on the number of tickets they bought. This means that the treatment assignment (D) is not completely independent from the potential outcomes (Y_i^1, Y_i^0). To satisfy the conditional exchangeability assumption,[5] our statistical model must control for the number of tickets (L) that each survey respondents bought and/or for all the Z covariates that link D to Y.

This regression model can be simple or flexible. It can be purely linear, or include a non-linear link function, multiplicative interactions, polynomials, or splines.[6] The model can satisfy the conditional exchangeability assumption using a minimal set of control variables (L), or it can include more covariates (Z) to improve precision.[7]

Here, we fit a linear regression where the treatment variable (`win_big`) is interacted with the number of tickets bought (`tickets`) and a series of demographic characteristics, including gender, age, employment, and pre-lottery labor earnings. To interact treatment and covariates, we insert a star * symbol and parentheses in the regression formula.

```
mod = lm(
  earnings_post_avg ~ win_big * (
    tickets + man + work + age + education + college + year +
    earnings_pre_1 + earnings_pre_2 + earnings_pre_3),
  data = dat)

summary(mod)
```

---

[5] In the Pearl (2009) tradition, we would say: "to satisfy the backdoor criterion."
[6] Chapter 10
[7] Section 9.1

	Estimate	Std. Error	t value	Pr(>|t|)
(Intercept)	1302.25641	1099.91390	1.184	0.2374
win_big	1327.56843	2435.94389	0.545	0.5862
tickets	0.13340	0.31632	0.422	0.6736
man	-0.85082	1.27346	-0.668	0.5046
work	3.24753	1.61780	2.007	0.0457
age	-0.23762	0.05330	-4.459	<0.001
education	-0.53448	0.53047	-1.008	0.3145
college	3.03770	2.08575	1.456	0.1464
year	-0.64566	0.55434	-1.165	0.2451
earnings_pre_1	0.05247	0.16390	0.320	0.7491
earnings_pre_2	-0.04408	0.18013	-0.245	0.8069
earnings_pre_3	0.86314	0.10288	8.390	<0.001
win_big:tickets	-0.17277	0.54143	-0.319	0.7499
win_big:man	1.51214	4.38811	0.345	0.7307
win_big:work	-0.73710	5.23892	-0.141	0.8882
win_big:age	0.13777	0.14432	0.955	0.3406
win_big:education	0.61646	1.11628	0.552	0.5812
win_big:college	9.62742	5.50191	1.750	0.0812
win_big:year	-0.67657	1.22789	-0.551	0.5821
win_big:earnings_pre_1	-0.23283	0.41698	-0.558	0.5770
win_big:earnings_pre_2	-0.26659	0.49941	-0.534	0.5939
win_big:earnings_pre_3	-0.12319	0.27092	-0.455	0.6497

8.1.4   Impute

The second G-computation step is to impute—or predict—the outcome for each observed individual, under different hypothetical treatment regimes. To do this, we create two identical copies of the dataset used to fit the model, and we use the `transform` function to fix the `win_big` variable to counterfactual values.

```
d0 = transform(dat, win_big = 0)
d1 = transform(dat, win_big = 1)
```

Using these two datasets, we predict the expected `earnings_post_avg` for each participant under counterfactual treatment regimes.

```
p0 = predictions(mod, newdata = d0)
p1 = predictions(mod, newdata = d1)
```

To take stock of what we have done so far, we can inspect the original dataset, and compare the predicted outcomes for an arbitrary survey respondent: the person in the sixth row of our dataset. By subsetting the data frame, we see that this person did *not*, in fact, win a big lottery prize.

```
dat[6, "win_big"]
```

```
[1] 0
```

Using our fitted model, we predicted that this person's average labor earnings—without lottery winnings—would be:

```
p0[6, "estimate"]
```

```
[1] 31.91696
```

If this individual belonged to the treatment group instead (`win_big=1`), our model would predict lower earnings:

```
p1[6, "estimate"]
```

```
[1] 24.5759
```

Thus, our model expects that, for an individual with these socio-demographic characteristics, winning the lottery is likely to decrease labor earnings. In the next section, we compare these predicted outcomes more systematically, at the aggregate level, to estimate the ATE, ATT, and ATU.

8.1.5   Compare

To estimate the treatment effect of `win_big` via G-computation, we aggregate individual-level predictions from the two counterfactual worlds. We can do this by extracting the vectors of estimates computed above and by calling the `mean()` function.

The average predicted outcome in the control counterfactual is:

```
mean(p0$estimate)
```

```
[1] 17.42834
```

The average predicted outcome in the treatment counterfactual is:

```
mean(p1$estimate)
```

```
[1] 11.86662
```

A more convenient way to obtain the same quantities is to call the `avg_predictions()` function from the **marginaleffects** package. As noted in Chapter 5, using the `variables` argument computes predicted values on a counterfactual grid, and the `by` argument marginalizes with respect to the focal variable. The results are the same as those we calculated manually, but they now come with standard errors and confidence intervals.

```
avg_predictions(mod,
  variables = "win_big",
  by = "win_big")
```

win_big	Estimate	Std. Error	z	Pr(>|z|)	2.5 %	97.5 %
0	17.4283	0.570124	30.56939	< 0.001	16.31092	18.5458
1	11.8666	2.278485	5.20812	< 0.001	7.40087	16.3324

The ATE is the difference between the two estimates printed above. We can compute it by subtraction, or use the `avg_comparisons()` function.

```
cmp = avg_comparisons(mod,
  variables = "win_big",
  newdata = dat)
cmp
```

Estimate	Std. Error	z	Pr(>|z|)	2.5 %	97.5 %
−5.56172	2.34873	−2.36797	0.0178861	−10.1652	−0.958293

This is the G-computation estimate of the average treatment effect of `win_big` on `earnings_post_avg`. On average, our model estimates that moving from the loser to the winner group decreases the predicted outcome by 5.6.

To compute the ATT, we use the same approach as above, but restrict our attention to the subset of individuals who were actually treated. We use the `subset()` function to select rows of the original dataset with treated individuals, and feed the resulting grid to the `newdata` argument.

```
avg_comparisons(mod,
  variables = "win_big",
  newdata = subset(win_big == 1))
```

Estimate	Std. Error	z	Pr(>|z|)	2.5 %	97.5 %
−9.48789	1.82433	−5.20076	< 0.001	−13.0635	−5.91227

The ATU is computed analogously, by restricting our attention to the subset of untreated individuals.

```
avg_comparisons(mod,
  variables = "win_big",
  newdata = subset(win_big == 0))
```

Estimate	Std. Error	z	Pr(>|z|)	2.5 %	97.5 %
−4.90989	2.59002	−1.8957	0.0580003	−9.98624	0.166455

It is interesting to notice that the three quantities of interest are quite different from one another. Indeed, the fact that the ATT is larger than the ATU suggests an asymmetry. Our model expects a strong decrease in earnings after winning a prize for people who are more likely to win (i.e., people who buy a lot of tickets). In contrast, our model expects a smaller decrease in earnings after winning a prize for people who are less likely to win (i.e., people who buy fewer tickets). This finding underscores the importance of clearly defining one's estimand before estimating and interpreting statistical models.[8]

Note that subtle issues arise when computing standard errors for G-computation estimates. In particular, standard errors calculated by assuming that covariates are fixed rather than sampled may not always have adequate coverage in the target population (Ding et al., 2023, ch.9). Many applied researchers have thus turned to the bootstrap as a simple strategy to quantify uncertainty around causal estimates (e.g., via the `inferences()` function from the `marginaleffects` package). Researchers have proposed analytic expressions for the unconditional variance (Ye et al., 2023; Hansen and Overgaard, 2024; Magirr et al., 2024). Interested readers can visit the marginaleffects.com website for a discussion of alternatives and implementation details.

8.2   Conditional treatment effects: CATE

The conditional average treatment effect (CATE) is an estimand that can be used to characterize how the effect of a treatment varies across subgroups. Unlike the ATE, which provides an overall average, the CATE allows us to see if certain individuals respond differently, depending on their individual characteristics. Formally, the CATE of D_i is written

$$E[Y_i^1 - Y_i^0 \mid X_i = x],$$

where Y_i^1 is the potential outcome when the treatment $D_i = 1$, and Y_i^0 is the potential outcome when the treatment $D_i = 0$. The $X_i = x$ after the vertical bar indicates that the average treatment effect is estimated separately for specific values of the covariate X_i, rather than averaged across the entire distribution of covariates. The conditioning variable X_i could represent discrete variables such as education or employment status.

To illustrate, let us estimate the effect of winning a big lottery prize on labor earnings, conditioning on employment status.

```
avg_comparisons(mod, variables = "win_big", by = "work")
```

---

[8]Section 2.2

work	Estimate	Std. Error	z	Pr($>$|z|)	2.5 %	97.5 %
0	−0.0708498	4.86727	−0.0145564	0.98838611	−9.61053	9.46883
1	−7.0973073	2.40030	−2.9568404	0.00310809	−11.80181	−2.39280

For individuals who were initially unemployed (`work=0`), the estimated average effect of winning a big lottery prize on labor earnings is −0.07 which is close to zero and not statistically significant. In contrast, for individuals who were employed (`work=1`), the estimated effect is -7.1, meaning that winning a big lottery prize significantly reduces labor earnings. These results highlight that the effect of a financial windfall is not uniform. While unemployed individuals do not adjust their earnings, those who were previously employed substantially reduce their labor income, likely by working fewer hours or leaving the labor force altogether. This pattern aligns with economic theories suggesting that an unexpected increase in wealth can reduce the incentive to work, but mostly for those with an existing attachment to the labor market.

It is useful to note that CATEs can also be estimated with additional conditioning variables. All we would have to do is add more categorical variables to the `by` argument of the `avg_comparisons()` function. Doing this would paint a more fine-grained portrait, but also reduce the number of observations (and information) available in each subgroup.

In conclusion, G-computation provides a useful framework for estimating causal effects from observational data, allowing researchers to draw meaningful inferences even in the absence of randomized experiments. By modeling, imputing, and comparing potential outcomes, we can target estimands such as the ATE, ATT, ATU, and CATE, while accounting for confounding variables. This chapter has demonstrated the application of G-computation using a lottery study, highlighting its utility in real-world scenarios.

9

Experiments

The analysis of experiments is a common use case for the `marginaleffects` package, which provides researchers with a powerful toolkit for estimating treatment effects, interpreting interactions, and visualizing results across experimental conditions. To illustrate these benefits, this chapter discusses two common applications: covariate adjustment, and the interpretation of results from a 2-by-2 factorial experiment. The marginaleffects.com website hosts more tutorials to guide the analysis and interpretation of a variety of other experimental designs.

9.1 Regression adjustment

The first application that we consider is a simple two-arm experiment, where participants are randomly assigned to a treatment or a control group. Our goal is to estimate the average treatment effect, while adjusting for covariates. As in previous chapters, we use data from Thornton (2008). The `outcome` is a binary variable that records if individuals traveled to the clinic to learn their HIV status. The treatment is a financial `incentive`, randomly given to participants, with the goal of encouraging testing.

Since the `incentive` treatment was randomized, we can estimate the average treatment effect (ATE) easily via linear regression on the binary outcome. This is the so-called linear probability model (LPM).

```
library(marginaleffects)
dat = get_dataset("thornton")
mod = lm(outcome ~ incentive, data = dat)
coef(mod)
```

```
(Intercept)    incentive
  0.3397746    0.4510603
```

As demonstrated in previous chapters, the `marginaleffects` package implements post-estimation transformations that are essentially agnostic with respect to the choice of model. Hence, the workflow described below would

DOI: 10.1201/9781003560333-9

remain unchanged if the analyst chose to fit a GLM (or other) model instead of this LPM.

The coefficient associated to the `incentive` variable shows the difference in the predicted probability of seeking one's test result, between those who received the incentive and those who did not. This result is identical to what we would obtain via G-computation, using the `avg_comparisons()` function.[1] One benefit of using this function instead of looking at individual coefficients, is that the `vcov` argument allows us to easily report heteroskedasticity-consistent, robust, or clustered standard errors.[2]

```
avg_comparisons(mod, variables = "incentive", vcov = "HC2")
```

Estimate	Std. Error	z	Pr(>|z|)	2.5 %	97.5 %
0.451	0.0209	21.6	<0.001	0.41	0.492

When the treatment is randomized, it is not strictly necessary to adjust for covariates in order to obtain an unbiased estimate of the average treatment effect. However, including control variables in the regression model can improve the precision of estimates, by reducing the unexplained variance in the outcome. This is an attractive strategy, but it must be applied with care. Indeed, as Freedman (2008) points out, naively inserting control variables in additive fashion to our model equation has an important downside: it can introduce small-sample bias and degrade asymptotic precision. To avoid this problem, Lin (2013) recommends a simple solution: interacting all covariates with the treatment indicator, and reporting heteroskedasticity-robust standard errors.

```
mod = lm(outcome ~ incentive * (age + distance + hiv2004), data = dat)
```

Interpreting the results of this model is not straightforward, since there are multiple coefficients, with several multiplicative interactions.[3]

```
coef(mod)
```

(Intercept)	incentive	age	distance
0.309537645	0.496482649	0.003495124	-0.042323583
hiv2004	incentive:age	incentive:distance	incentive:hiv2004
0.013562711	-0.002177778	0.014571883	-0.070766068

---

[1]Chapter 8

[2]Section 14.1.3

[3]Lin (2013) recommends interacting the treatment indicator with de-meaned control variables. By doing so, the linear regression coefficient of the treatment variable become directly interpretable. Using the `avg_comparisons()` function is more convenient, because it produces the same results without forcing us to de-mean control variables first. As Lin notes, it may still be a good idea to report the unadjusted estimate, for transparency reasons, and to avoid the temptation of specification-searching.

Thankfully, we can simply call `avg_comparisons()` again to get the covariate-adjusted estimate, along with robust standard errors.

```
avg_comparisons(mod,
    variables = "incentive",
    vcov = "HC2")
```

Estimate	Std. Error	z	Pr(>|z|)	2.5 %	97.5 %
0.449	0.0209	21.5	<0.001	0.408	0.49

This suggests that, on average, receiving a monetary `incentive` increases by 44.9 percentage points the probability of seeking one's HIV test result.

9.2   Factorial experiments

A factorial experiment is a type of study design that allows researchers to assess the effects of two or more randomized treatments simultaneously, with each treatment having multiple levels or conditions. A common example is the 2-by-2 design, where two variables with two levels are randomized simultaneously and independently. This setup enables the evaluator to quantify the effects of each treatment, but also to examine if the treatments interact with one another.

In medicine, factorial experiments are widely used to explore complex phenomena like drug interactions, where researchers test different combinations of treatments to see how they affect patient outcomes. In plant physiology, they could be used to see how combinations of temperature and humidity affect photosynthesis. In a business context, a company could use a factorial design to test if different advertising strategies and price points drive sales.

To illustrate, let's consider a simple data set with 32 observations, a numeric outcome Y, and two treatments with two levels: $T_a = \{0, 1\}$ and $T_b = \{0, 1\}$. We fit a linear regression model with each treatment entered individually, along with their multiplicative interaction.[4]

```
library(marginaleffects)
dat = get_dataset("factorial_01")
mod = lm(Y ~ Ta + Tb + Ta:Tb, data = dat)
coef(mod)
```

```
(Intercept)          Ta          Tb       Ta:Tb
  15.050000    5.692857    4.700000    2.928571
```

---
[4]Chapter 10

We can plot the model's predictions for each combination of T_a and T_b using the `plot_predictions` function.

```
plot_predictions(mod, by = c("Ta", "Tb"))
```

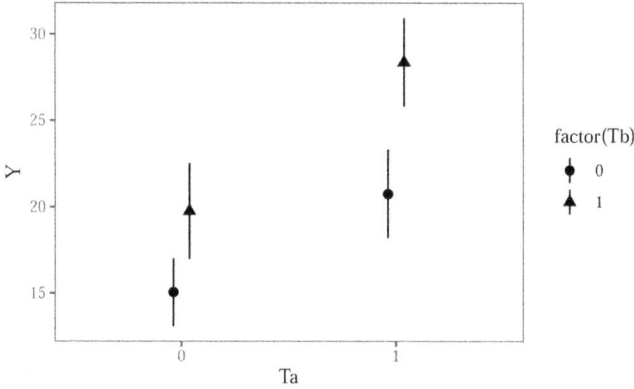

Now, imagine we want to estimate the effect of changing T_a from 0 to 1 on Y, while holding $T_b = 0$. We define two counterfactual values of Y and take their difference.

$$\hat{Y} = \hat{\beta}_1 + \hat{\beta}_2 \cdot T_a + \hat{\beta}_3 \cdot T_b + \hat{\beta}_4 \cdot T_a \cdot T_b$$
$$\hat{Y}_{T_a=0, T_b=0} = \hat{\beta}_1 + \hat{\beta}_2 \cdot 0 + \hat{\beta}_3 \cdot 0 + \hat{\beta}_4 \cdot 0 \cdot 0$$
$$\hat{Y}_{T_a=1, T_b=0} = \hat{\beta}_1 + \hat{\beta}_2 \cdot 1 + \hat{\beta}_3 \cdot 0 + \hat{\beta}_4 \cdot 1 \cdot 0$$
$$\hat{Y}_{T_a=1, T_b=0} - \hat{Y}_{T_a=0, T_b=0} = \hat{\beta}_2 = 5.693$$

We can use a similar approach to estimate a cross-contrast, that is, the effect of changing both T_a and T_b from 0 to 1 simultaneously.

$$\hat{Y}_{T_a=0, T_b=0} = \hat{\beta}_1 + \hat{\beta}_2 \cdot 0 + \hat{\beta}_3 \cdot 0 + \hat{\beta}_4 \cdot 0 \cdot 0$$
$$\hat{Y}_{T_a=1, T_b=1} = \hat{\beta}_1 + \hat{\beta}_2 \cdot 1 + \hat{\beta}_3 \cdot 1 + \hat{\beta}_4 \cdot 1 \cdot 1$$
$$\hat{Y}_{T_a=1, T_b=1} - \hat{Y}_{T_0=1, T_b=0} = \hat{\beta}_2 + \hat{\beta}_3 + \hat{\beta}_4 = 13.321$$

In a simple 2-by-2 experiment, this arithmetic is fairly simple. However, things can quickly get complicated when more treatment conditions are introduced, or when regression models include non-linear components. Moreover, even if we can take sums of coefficients to get the estimates we care about, computing standard errors (classical or robust) is non-trivial. Thus, it is convenient to use software like **marginaleffects** to do the hard work for us.

To estimate the effect of T_a on Y, holding $T_b = 0$ constant, we use the `avg_comparisons()` function. The `variables` argument specifies the treatment of interest, and the `newdata` argument specifies the levels of the other treatment(s) at which we want to evaluate the contrast. In this case, we use `newdata=subset(Tb == 0)` to compute the quantity of interest only in the subset of observations where $T_b = 0$.

```
avg_comparisons(mod,
  variables = "Ta",
  newdata = subset(Tb == 0))
```

Estimate	Std. Error	z	Pr(>|z|)	2.5 %	97.5 %
5.69	1.65	3.45	<0.001	2.46	8.93

To estimate a cross-contrast, we specify each treatment of interest in the `variables` argument, and set `cross=TRUE`. This gives the estimated effect of simultaneously moving T_a and T_b from 0 to 1.

```
avg_comparisons(mod,
  variables = c("Ta", "Tb"),
  cross = TRUE)
```

C: Ta	C: Tb	Estimate	Std. Error	z	Pr(>|z|)	2.5 %	97.5 %
1 − 0	1 − 0	13.3	1.65	8.07	<0.001	10.1	16.6

Figure 9.1 shows six contrasts of interest in a 2-by-2 experimental design. The vertical axis shows the value of the outcome variable in each treatment group. The horizontal axis shows the value of the first treatment T_a. The shape of points—0 or 1—represents the value of the T_b treatment. For example, the "1" symbol in the top left shows the average value of Y in the $T_a = 0, T_b = 1$ group.

In each panel, the arrow links two treatment groups that we want to compare. For example, in the top-left panel we compare $\hat{Y}_{T_a=0,T_b=0}$ to $\hat{Y}_{T_a=1,T_b=0}$. In the bottom-right panel, we compare $\hat{Y}_{T_a=0,T_b=1}$ to $\hat{Y}_{T_a=1,T_b=0}$. The `avg_comparisons()` call at the top of each panel shows how to use `marginaleffects` to compute the difference in Y that results from the illustrated change in treatment conditions.

One reason factorial experiments are so popular is that they allow researchers to assess the possibility that treatments interact with one another. Indeed, in some cases, the effect of a treatment T_a will be stronger (or weaker) at different values of treatment T_b.

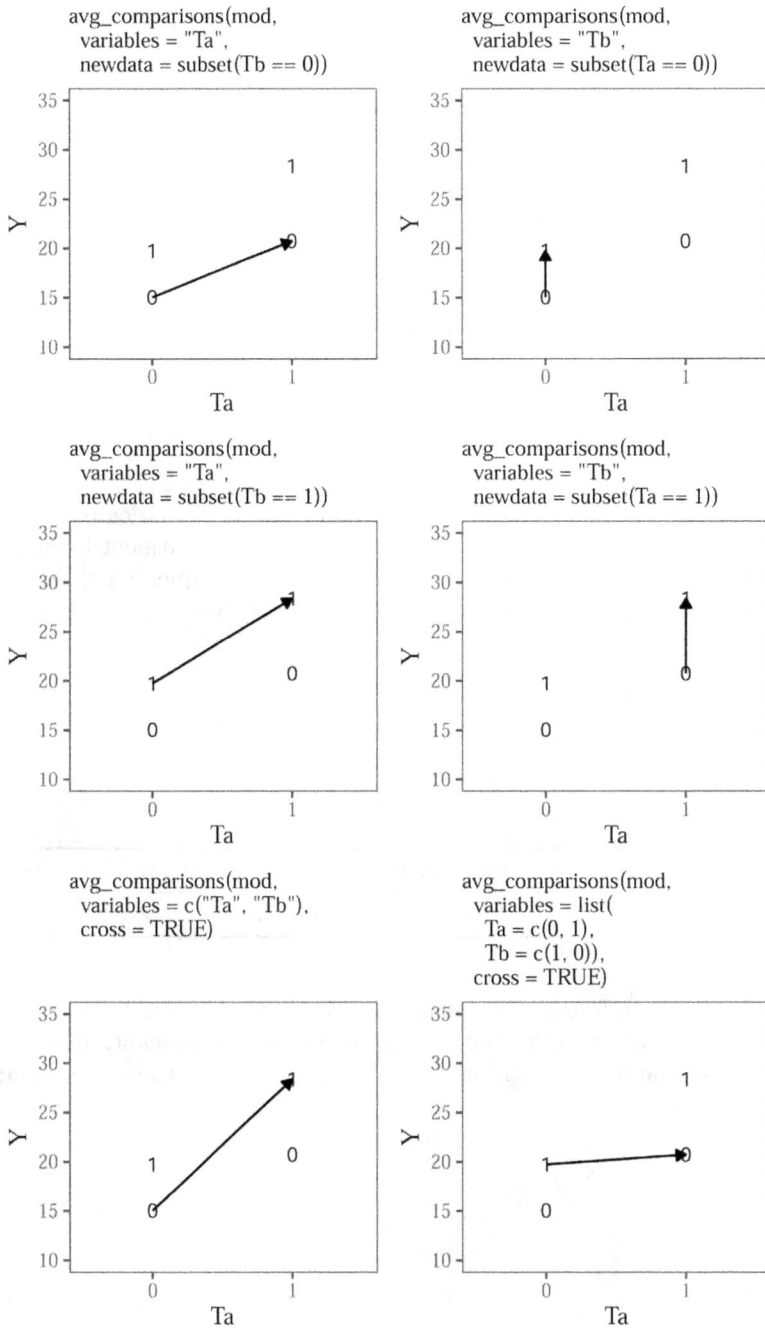

Figure 9.1: Six contrasts of interest in a 2-by-2 experimental design. The x-axis shows values of the T_a treatment and the 0/1 symbols on the plotting surface indicate values of the T_b treatment.

To check this, let us estimate the effect of a change from 0 to 1 in T_a, in the subsets of observations with different values of T_b.

```
cmp <- avg_comparisons(mod, variables = "Ta", by = "Tb")
cmp
```

Tb	Estimate	Std. Error	z	Pr(>|z|)	2.5 %	97.5 %
0	5.69	1.65	3.45	<0.001	2.46	8.93
1	8.62	1.93	4.46	<0.001	4.84	12.41

The estimated effect of T_a on Y is equal to 5.69 when $T_b = 0$, but equal to 8.62 when $T_b = 1$. At first glance, it thus seems that the effect of T_a is stronger when $T_b = 1$.

Is the difference between estimated treatment effects statistically significant? Does T_b moderate the effect the effect of T_a? Is there treatment heterogeneity or interaction between T_a and T_b? To answer these questions, we use the **hypothesis** argument and check if the difference between subgroup estimates is significant.

```
avg_comparisons(mod,
  variables = "Ta",
  by = "Tb",
  hypothesis = "b2 - b1 = 0")
```

Hypothesis	Estimate	Std. Error	z	Pr(>|z|)	2.5 %	97.5 %
b2-b1=0	2.93	2.54	1.15	0.249	−2.05	7.91

While there is a difference between our two estimates, the z statistic and p value indicate that this difference is not statistically significant. Therefore, we cannot reject the null hypothesis that the effect of T_a is the same, regardless of the value of T_b.

10

Interactions and polynomials

So far, the models that we have considered were relatively simple. In this chapter, we apply the same workflow, framework, and software introduced in Parts I and II of the book to interpret estimates from slightly more complex specifications. Our goal is to address two related issues: heterogeneity and flexibility.

We say that there is "heterogeneity" when the association between an independent variable and a dependent variable varies across contexts, or based on the characteristics of our objects of study. For instance, a new medical treatment might significantly reduce blood pressure in younger adults, but have a weaker effect on older ones. Or a marketing campaign may increase sales in rural areas but not urban ones. Applied statisticians use different expressions to describe such situations: heterogeneity, moderation, interaction, effect modification, context-conditionality, or strata-specific effects. In this chapter, we use these expressions interchangeably to describe situations where the strength of association between two variables depends on a third: the moderator.[1]

The second issue that this chapter tackles is flexibility. Many of the statistical models that data analysts fit are relatively simple and rigid; they often rely on linear approximations of the data generating process. Such approximations may not be appropriate when the relationships of interest are complex or non-linear.[2] In environmental science, for example, the relationship between temperature and crop yield is often non-linear: as temperatures rise, crop yields may initially increase due to optimal growing conditions, but eventually decline as heat stress becomes detrimental.

---

[1]The term "moderation" does not necessarily imply that the moderator M *causes* the relationship between X and Y to change. For example, imagine that a public health intervention is shown to have stronger effects in some specific cities. An organization with limited resources may wish to deploy the intervention where it is most effective, even if the organization does not understand exactly *why* the intervention is more effective in those cities. To learn more about the difference between "treatment effect heterogeneity" and "causal moderation analysis," and about the more stringent assumptions required for the latter, see Bansak (2021) and VanderWeele (2009).

[2]Two downsides of more flexible models are that they can sometimes reduce statistical power or overfit the data.

Complex relationships and heterogeneous effects are present in virtually all empirical domains. This chapter shows that the conceptual framework and software tools presented in earlier parts of this book can be applied in straightforward fashion to gain insights into complex phenomena. We will focus on two modeling strategies to account for heterogeneity and increase the flexibility of our models: multiplicative interactions and polynomials.

These two approaches do not exhaust the set of techniques that researchers may want to deploy to study complex phenomena, but the same ideas and workflows can be applied to interpret even more flexible models. Readers who wish to go further in that direction can turn to Chapter 13 on Machine Learning or visit marginaleffects.com for tutorials on Generalized Additive Models, splines, and more.

10.1   Multiplicative interactions

We say that there is heterogeneity when the strength of the association between an explanator X and an outcome Y varies based on the value of a moderator M. The association between X and Y can be stronger, weaker, or completely reversed for different values of M. Typically, M is a variable which measures contextual elements, or features of the individuals, groups, or units under observation.

One common strategy to study moderation is to fit a statistical model with multiplicative interactions (Brambor et al., 2006; Kam and Franzese, Jr., 2009; Clark and Golder, 2023). This involves creating a new composite variable by multiplying the explanator (X) to the moderator (M). Then, we insert the interacted variable as a predictor in the model, alongside its individual components.[3] One popular specification for this type of moderation analysis is a simple linear model with one interaction.

$$Y = \beta_1 + \beta_2 \cdot X + \beta_3 \cdot M + \beta_4 \cdot X \cdot M + \varepsilon, \qquad (10.1)$$

where Y is the outcome, X is the predictor of interest, and M is a contextual variable which moderates the relationship between X and Y.

To characterize the association between X, M, and Y, we can plug the values of these variables into Equation 10.1. When $M = 0$, moving from $X = 0$ to $X = 1$ is associated with an increase of β_2 in predicted Y.

---

[3]In most cases, it is important to include all constitutive terms in addition to interactions. For example, if a model includes a multiplication between three variables $X \cdot W \cdot Z$, one would typically want to also include $X \cdot W, X \cdot Z, W \cdot Z, X, W$, and Z. See Clark and Golder (2023) for details.

$$Y_{X=1,M=0} = \beta_1 + \beta_2 \cdot 1 + \beta_3 \cdot 0 + \beta_4 \cdot 1 \cdot 0 + \varepsilon$$
$$Y_{X=0,M=0} = \beta_1 + \beta_2 \cdot 0 + \beta_3 \cdot 0 + \beta_4 \cdot 0 \cdot 0 + \varepsilon$$
$$Y_{X=1,M=0} - Y_{X=0,M=0} = \beta_2$$

In contrast, when $M = 1$, a change from $X = 0$ to $X = 1$ is associated with a change of $\beta_2 + \beta_4$ in predicted Y.

$$Y_{X=1,M=1} = \beta_1 + \beta_2 \cdot 1 + \beta_3 \cdot 1 + \beta_4 \cdot 1 \cdot 1 + \varepsilon$$
$$Y_{X=0,M=1} = \beta_1 + \beta_2 \cdot 0 + \beta_3 \cdot 1 + \beta_4 \cdot 0 \cdot 1 + \varepsilon$$
$$Y_{X=1,M=1} - Y_{X=0,M=1} = \beta_2 + \beta_4$$

The difference between these two results shows what we mean by heterogeneity, moderation, or effect modification: the strength of association between a dependent and an independent variable varies based on the value of a moderator.

The parameters in Equation 10.1 are fairly straightforward to grasp, and we can interpret them using a little algebra. However, as soon as a model includes non-linear components or more than one interaction, it quickly becomes very difficult to interpret coefficient estimates directly. This book has argued that it is usually best to ignore raw coefficients, and to focus on more interpretable quantities, like predictions, counterfactual comparisons, and slopes. We will follow the same advice here.

When models include interactions, the presentation of results can depend on the nature of X and M. In what follows, we consider different scenarios where the predictor and moderator are categorical or numeric variables.

10.1.1 Categorical-by-categorical

The first case to consider is when the association between a categorical explanator X and an outcome Y is moderated by a categorical variable M. For example, this could occur if the effect of a drug (X) on patient recovery (Y) varies across different age groups (M).

Let's consider a simulated dataset with three variables: the outcome Y and the treatment X are binary variables, and the moderator M is a categorical variable with 3 levels (a, b, and c).

```
library(marginaleffects)
library(ggplot2)
library(patchwork)
dat = get_dataset("interaction_01")
head(dat)
```

```
# A data frame: 6 x 3
       Y     X M
* <int> <int> <chr>
1     1     0 a
2     1     1 b
3     1     0 a
4     0     0 a
5     1     0 c
6     1     1 a
```

We fit a logistic regression model using the `glm()` function. This model includes X and M as standalone predictors, and it also includes a multiplicative interaction between those two variables. To create this interaction, we insert a colon (`:`) character in the model formula.

```
mod = glm(Y ~ X + M + X:M, data = dat, family = binomial)
```

Alternatively, one could use the asterisk (`*`) shortcut, which automatically includes the multiplicative interaction, as well as each of the individual variables. The next command is thus equivalent to the previous one.

```
mod = glm(Y ~ X * M, data = dat, family = binomial)
```

As is the case for many of the more complex models that we will consider, the coefficient estimates for this logit model with interactions are difficult to interpret on their own.

```
summary(mod)
```

```
Coefficients:
              Estimate Std. Error z value Pr(>|z|)
(Intercept)    0.43643    0.07195   6.065   <0.001
X              0.26039    0.10337   2.519   0.0118
Mb             0.56596    0.10699   5.290   <0.001
Mc             0.96967    0.11087   8.746   <0.001
X:Mb           0.89492    0.17300   5.173   <0.001
X:Mc           1.41219    0.21462   6.580   <0.001
```

Thankfully, we can rely on the framework and tools introduced in Parts I and II of this book to make these results intelligible.

Marginal predictions

A good first step toward interpretation is to compute the average predicted outcome for each combination of predictor X and moderator M. We can do this by calling the `avg_predictions()` function with its by argument. As explained in Section 5.3, this is equivalent to computing fitted values for every row in the original data, and then taking the average of fitted values by subgroup.[4]

```
avg_predictions(mod, by = c("X", "M"))
```

X	M	Estimate	Std. Error	z	Pr($>$|z|)	2.5 %	97.5 %
0	a	0.607	0.01716	35.4	<0.001	0.574	0.641
0	b	0.732	0.01555	47.0	<0.001	0.701	0.762
0	c	0.803	0.01334	60.2	<0.001	0.777	0.829
1	a	0.667	0.01647	40.5	<0.001	0.635	0.700
1	b	0.896	0.01058	84.7	<0.001	0.876	0.917
1	c	0.956	0.00707	135.2	<0.001	0.942	0.970

When $X = 0$ and $M = a$, the average prediction is 0.607. Since the outcome is a binary variable, predicted values for this specific model are expressed on the $[0, 1]$ interval, and they can be interpreted as probabilities.

There appears to be considerable variation in the average predicted probability that $Y = 1$ across subgroups, going from 61% to 96%. This stark variation can be underscored visually using the `plot_predictions()` function.

```
plot_predictions(mod, by = c("M", "X"))
```

The triangles in this figure are systematically above the corresponding circles. On average, the predicted probability that $Y = 1$ is considerably higher when $X = 1$ than when $X = 0$. Figure 10.1 also gives us a first hint that the relationship between X and Y may differ across values of M. For some values of M, the triangles are farther apart from the circles, suggesting that the effect of X on Y may be stronger when M takes on those values.

Does X affect Y ?

We can build on these preliminary findings by adopting a more explicitly counterfactual approach, using the `comparisons()` family of function.[5] Recall, from Chapter 6, that we can compute an average counterfactual comparison by taking three steps.

---

[4]An alternative would be to report "marginal means" by computing predictions on a balanced grid before aggregating. This can be achieved by calling: `avg_predictions(mod, newdata="balanced", by=c("X","M"))`

[5]For a discussion of the difference between hypothesis tests conducted on average predictions and average comparisons, see Section 6.3.1.

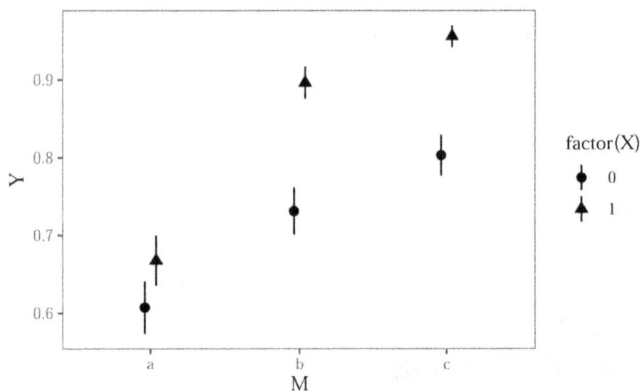

Figure 10.1: Average predicted probability that Y=1 for different values of M and X.

1. Modify the original dataset to fix $X = 0$ for all observations, and compute predictions for every row.
2. Modify the original dataset to fix $X = 1$ for all observations, and compute predictions for every row.
3. Calculate the average difference between the counterfactual predictions computed in steps 1 and 2.

This can be done with a single line of code.

```
avg_comparisons(mod, variables = "X")
```

Estimate	Std. Error	z	Pr(>|z|)	2.5 %	97.5 %
0.127	0.0112	11.3	<0.001	0.105	0.149

On average, moving from 0 to 1 on the X variable is associated with an increase of 0.127 on the outcome scale. Since we fit a logistic regression model, predictions are expressed on the probability scale. Thus, the estimate printed above suggests that the average treatment effect of X on Y is about 12.7 percentage points.[6] This estimate is statistically distinguishable from zero, as the small p value and confidence interval attest.

---

[6]Interpreting this quantity in causal terms, as an average treatment effect, assumes that there is no confounding. See Chapter 8.

Is the effect of X on Y moderated by M?

Now we can exploit the multiplicative interaction in our model to interrogate heterogeneity. To see if the effect of X on Y depends on M, we make the same function call as above, but add the by argument.

```
avg_comparisons(mod, variables = "X", by = "M")
```

M	Estimate	Std. Error	z	Pr(>|z|)	2.5 %	97.5 %
a	0.0601	0.0238	2.53	0.0115	0.0135	0.107
b	0.1649	0.0188	8.77	<0.001	0.1280	0.202
c	0.1529	0.0151	10.13	<0.001	0.1233	0.182

On average, moving from the control ($X = 0$) to the treatment group ($X = 1$) is associated with an increase of 6 percentage points for individuals in category a. The average estimated effect of X for individuals in category c is 15 percentage points.

At first glance, these two estimates look different. But is the difference between 6 and 15 percentage points statistically significant? Can we reject the null hypothesis that these two estimates are equal? To answer these questions, we can use the hypothesis argument.

```
avg_comparisons(mod,
  variables = "X",
  by = "M",
  hypothesis = "b3 - b1 = 0")
```

Hypothesis	Estimate	Std. Error	z	Pr(>|z|)	2.5 %	97.5 %
b3-b1=0	0.0928	0.0282	3.29	<0.001	0.0376	0.148

The difference between the average estimated effect of X in categories c and a is: $0.1529 - 0.0601 = 0.0928$. This difference is associated to a large z statistic and a small p value. Therefore, we can conclude that the difference is statistically significant; we can reject the null hypothesis that the effect of X on Y is the same in sub-populations a and c.

10.1.2   Categorical-by-continuous

The second case to consider is an interaction between a categorical explanator
(X) and a continuous mediator (M). To illustrate, we consider a new simulated
dataset.

```
dat = get_dataset("interaction_02")
head(dat)
```

```
# A data frame: 6 x 3
       Y      X       M
* <int> <int>   <dbl>
1      1     0  -0.521
2      0     1   0.136
3      1     0  -0.412
4      0     0  -0.567
5      0     0  -0.979
6      1     1  -1.13
```

As before, we fit a logistic regression model with a multiplicative interaction,
using the asterisk operator.

```
mod = glm(Y ~ X * M, data = dat, family = binomial)
summary(mod)
```

```
Coefficients:
              Estimate Std. Error z value Pr(>|z|)
(Intercept)  -0.83981    0.04644 -18.083  < 0.001
X             0.19222    0.06334   3.035  0.00241
M            -0.75891    0.05049 -15.030  < 0.001
X:M           0.35209    0.06753   5.213  < 0.001
```

Conditional predictions

In the previous section, we started by computing average predictions for each
combination of the interacted variable. When one of the variables is continuous
and takes on many values (like M), it is not practical to report averages for
every combination of X and M. Therefore, we focus on "conditional" estimates,
obtained by calling the **predictions()** function. We use the **datagrid()** and
fivenum() functions to create a grid of predictors based on Tukey's five number
summary of M.[7]

```
predictions(mod, newdata = datagrid(X = c(0, 1), M = fivenum))
```

---

[7]These five numbers correspond to elements of a standard boxplot: minimum, lower-hinge,
median, upper-hinge, and maximum.

X	M	Estimate	Pr(>|z|)	2.5 %	97.5 %
0	−3.14813	0.8248	<0.001	0.7766	0.8645
0	−0.65902	0.4159	<0.001	0.3921	0.4401
0	0.00407	0.3009	<0.001	0.2821	0.3204
0	0.68359	0.2045	<0.001	0.1848	0.2256
0	3.53494	0.0287	<0.001	0.0198	0.0414
1	−3.14813	0.6532	<0.001	0.5871	0.7139
1	−0.65902	0.4063	<0.001	0.3830	0.4300
1	0.00407	0.3432	<0.001	0.3244	0.3624
1	0.68359	0.2838	<0.001	0.2623	0.3063
1	3.53494	0.1105	<0.001	0.0820	0.1473

The results show considerable variation in the predicted probability that Y equals 1, ranging from 0.029 to 0.825.

Instead of making predictions for discrete values of the continuous moderator M, we can also draw a plot with that variable on the x-axis.

```
plot_predictions(mod, condition = c("M", "X"))
```

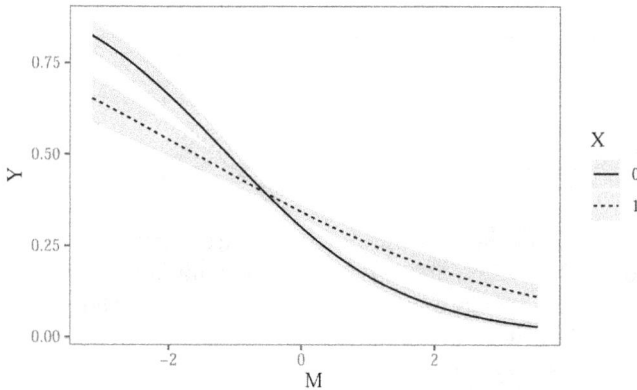

Figure 10.2: Predicted probability that Y=1, for different values of X and M.

Figure 10.2 shows that predicted values of Y tend to be lower when M is large. That figure also suggests that the relationship between X and Y has a different character for different values of M. When M is small (left side of the plot), we see

$$P(Y = 1|X = 1, M) < P(Y = 1|X = 0, M) \qquad \text{for small values of } M.$$

When M is large, the converse seems true.

Does X affect Y?

Moving to the counterfactual analysis, we call `avg_comparisons()` to get an overall estimate of the effect of X on the predicted $Pr(Y = 1)$.

```
avg_comparisons(mod, variables = "X")
```

Estimate	Std. Error	z	Pr(>|z|)	2.5 %	97.5 %
0.0284	0.0129	2.21	0.0273	0.00318	0.0536

On average, moving from 0 to 1 on X increases the predicted probability that $Y = 1$ by 2.8 percentage points.

Is the effect of X on Y moderated by M?

As explained in Chapter 6, we can estimate the effect of X for different values of M by using the **newdata** argument and **datagrid()** function. Here, we measure the strength of association between X and Y for two different values of M: its minimum and maximum.

```
comparisons(mod, variables = "X", newdata = datagrid(M = range))
```

M	Estimate	Std. Error	z	Pr(>|z|)	2.5 %	97.5 %
−3.15	−0.1716	0.0395	−4.35	<0.001	−0.2489	−0.0943
3.53	0.0818	0.0174	4.70	<0.001	0.0477	0.1159

Moving from 0 to 1 on the X variable is associated with a change of -0.172 in the predicted Y when the moderator M is at its minimum. Moving from 0 to 1 on the X variable is associated with a change of 0.082 in the predicted Y when the moderator M is at its maximum. Both of these estimates are associated with small p values, so we can reject the null hypotheses that they are equal to zero.

Both estimates are different from zero, but are they different from one another? Is the effect of X on Y different when M takes on different values? To check this, we can add the **hypothesis** argument to the previous call.

```
comparisons(mod,
  hypothesis = "b2 - b1 = 0",
  variables = "X",
  newdata = datagrid(M = range))
```

Hypothesis	Estimate	Std. Error	z	Pr(>|z|)	2.5 %	97.5 %
b2-b1=0	0.253	0.0546	4.64	<0.001	0.146	0.36

This confirms that the estimates are statistically distinguishable. We can reject the null hypothesis that M has no moderating effect on the relationship between X and Y.

10.1.3 Continuous-by-continuous

The third case to consider is an interaction between two continuous numeric variables: X and M. To illustrate, we fit a new model to simulated data.

```
dat = get_dataset("interaction_03")
head(dat)

# A data frame: 6 x 3
      Y      X       M
* <int>  <dbl>   <dbl>
1     0 -0.779  -0.761
2     0 -0.389  -0.199
3     0 -2.03    1.46
4     1 -0.982  -0.0254
5     1  0.248  -0.664
6     0 -2.10   -0.314

mod = glm(Y ~ X * M, data = dat, family = binomial)
summary(mod)

Coefficients:
            Estimate Std. Error z value Pr(>|z|)
(Intercept) -0.95119    0.03320 -28.650   <0.001
X            0.44657    0.03470  12.868   <0.001
M            0.21494    0.03435   6.257   <0.001
X:M          0.50298    0.03660  13.744   <0.001
```

Conditional predictions

As in the previous cases, we begin by computing the predicted outcomes for different values of the predictors. In practice, the analyst should report predictions for predictor values that are meaningful to the domain of application. Here, we hold X and M to fixed arbitrary values:

```
p = predictions(mod, newdata = datagrid(
  X = c(-2, 2),
  M = c(-1, 0, 1)))
p
```

X	M	Estimate	Pr(>|z|)	2.5 %	97.5 %
−2	−1	0.2586	<0.001	0.2215	0.2995
−2	0	0.1365	<0.001	0.1189	0.1563
−2	1	0.0669	<0.001	0.0531	0.0839
2	−1	0.2177	<0.001	0.1837	0.2561
2	0	0.4855	0.425	0.4500	0.5212
2	1	0.7619	<0.001	0.7213	0.7982

When $X = -2$ and $M = -1$, the predicted probability that Y equals 1 is 25.86%.

Rather than focus on arbitrary values of X and M, we can plot predicted values to communicate a richer set of estimates. When calling `plot_predictions()` with these data, we obtain a plot of predicted outcomes with the focal variable on the x-axis, and lines representing different values of the moderator.

```
plot_predictions(mod, condition = c("X", "M"))
```

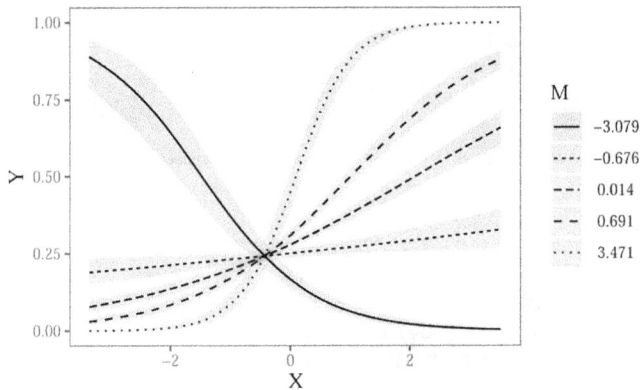

Figure 10.3: Predicted value of Y for different values of X and M.

We can draw two preliminary conclusions from Figure 10.3. First, the predicted values of Y depend strongly on the value of X. Moving from left to right in the plot often has a strong effect on the heights of predicted probability curves. Second, M strongly moderates the relationship between X and Y. Indeed, for some values of M the relationship of interest completely flips. For example, when M is around -3, the relationship between X and Y is negative: an increase in X is associated with a decrease in $Pr(Y = 1)$. However, for all other displayed values of M, the relationship between X and Y is positive: an increase in X is associated with an increase in $Pr(Y = 1)$.

Does X affect Y?

To measure the association between X on Y, we have two basic options. First, we can use the comparisons() function to compute the estimated effect of a discrete change in X on the predicted value of Y. Second, we can use the slopes() function to compute the estimated slope of Y with respect to X. Since the focal variable and mediator are both continuous, the rest of this section focuses on slopes. However, it is useful to note that slopes and counterfactual comparisons both offer valid answers to the research question, both families of functions operate in nearly identical ways, and both sets of results would lead us to the same substantive conclusions.

```
s = avg_slopes(mod, variables = "X")
s
```

Estimate	Std. Error	z	Pr(>|z|)	2.5 %	97.5 %
0.0857	0.00571	15	<0.001	0.0745	0.0969

On average, across all observed values of the moderator M, increasing X by one unit increases the predicted outcome by 0.086.[8] This is interesting but, as Figure 10.3 suggests, there is strong heterogeneity in the relationship of interest. The association between X and Y may be very different for different values of M.

Is the effect of X on Y moderated by M?

To answer this question, we estimate slopes of Y with respect to X, for five values of the moderator M.

```
slopes(mod, variables = "X", newdata = datagrid(M = fivenum))
```

M	Estimate	Std. Error	z	Pr(>|z|)	2.5 %	97.5 %
−3.0788	−0.1525	0.01899	−8.03	< 0.001	−0.18968	−0.1153
−0.6759	0.0200	0.00756	2.65	0.00815	0.00519	0.0348
0.0137	0.0913	0.00695	13.14	< 0.001	0.07768	0.1049
0.6906	0.1698	0.00950	17.87	< 0.001	0.15113	0.1884
3.4711	0.5426	0.03363	16.14	< 0.001	0.47673	0.6085

The results in this table confirm our intuition. When $M \approx -3$, the slope is negative: increasing X results in a reduction of Y. However, for the other

---

[8]Recall from Chapter 7 that this interpretation is valid for small changes in the neighborhoods where slopes are evaluated.

values of M, the average slope is positive. This is consistent with Figure 10.3, which shows one line with downward slope, and four lines with upward slopes.

For a more fine-grained analysis, we can plot the slope of Y with respect to X for all observed values of the moderator M.

```
plot_slopes(mod, variables = "X", condition = "M") +
    geom_hline(yintercept = 0, linetype = "dotted")
```

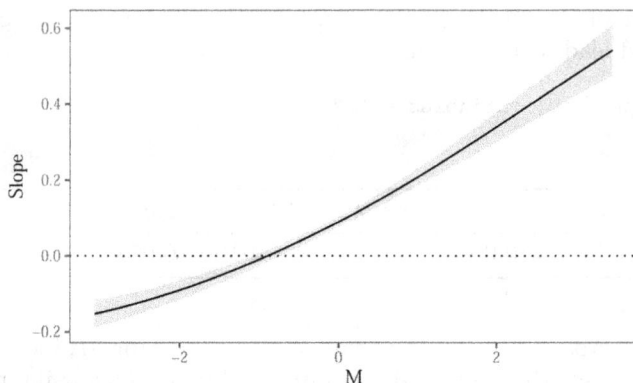

Figure 10.4: Slope of Y with respect to X, for different values of the moderator M.

Figure 10.4 shows that when the moderator M is below -1, the relationship between X and Y is negative: increasing X decreases Y. However, when M rises above -1, the relationship between X and Y becomes positive: increasing X increases Y.

We can confirm that this moderation effect is statistically significant using the **hypothesis** argument. First, we estimate the slope of Y with respect to X for two different values of the moderator M: its minimum and maximum.

```
slopes(mod,
  variables = "X",
  newdata = datagrid(M = range))
```

M	Estimate	Std. Error	z	Pr($>$|z|)	2.5 %	97.5 %
-3.08	-0.152	0.0190	-8.03	<0.001	-0.190	-0.115
3.47	0.543	0.0336	16.14	<0.001	0.477	0.609

Then, we use the **hypothesis** argument to compare the two estimates.

```
slopes(mod,
  variables = "X",
  newdata = datagrid(M = range),
  hypothesis = "b2 - b1 = 0")
```

Hypothesis	Estimate	Std. Error	z	Pr(>|z|)	2.5 %	97.5 %
b2-b1=0	0.695	0.0475	14.6	<0.001	0.602	0.788

The $\frac{\partial Y}{\partial X}$ slope is larger when evaluated at maximum M, than at minimum M. Therefore, we can reject the null hypothesis that M has no moderating effect on the relationship between X and Y.

10.1.4   Multiple interactions

The fourth case to consider is when more than two variables are included in multiplicative interactions. Such models have serious downsides: they can overfit the data, and they impose major costs in terms of statistical power, typically requiring considerably larger sample sizes than models without interaction. On the upside, models with multiple interactions allow more flexibility in modeling, and they can capture complex patterns of moderation between regressors.

Models with several multiplicative interactions do not pose any particular interpretation challenge, since the tools and workflows introduced in this book can be applied to these models in straightforward fashion. Consider this simulated dataset with four binary variables.

```
dat = get_dataset("interaction_04")
head(dat)
```

```
# A data frame: 6 x 4
        Y       X      M1      M2
* <int>  <int>  <int>  <int>
1       0       0       0       0
2       1       1       0       1
3       0       0       1       0
4       1       0       1       0
5       0       0       0       1
6       0       1       0       1
```

We fit a logistic regression model with Y as the outcome, and multiplicative interactions between all three predictors.

```
mod = glm(Y ~ X * M1 * M2, data = dat, family = binomial)
summary(mod)
```

```
Coefficients:
              Estimate Std. Error z value Pr(>|z|)
(Intercept)    -1.0209     0.0908 -11.244  <0.001
X               0.4632     0.1229   3.768  <0.001
M1             -0.7954     0.1470  -5.411  <0.001
M2              0.5788     0.1217   4.755  <0.001
X:M1            0.6746     0.1890   3.569  <0.001
X:M2           -0.9649     0.1716  -5.623  <0.001
M1:M2           0.4046     0.1890   2.141  0.0323
X:M1:M2         0.1976     0.2542   0.777  0.4369
```

Once again, the coefficient estimates of this logistic regression are difficult to interpret on their own, so we use functions from the `marginaleffects` package.

Marginal predictions

As before, we can compute and display marginal predicted outcomes in any subgroup of interest, using the `avg_predictions()` or `plot_predictions()` functions.

```
plot_predictions(mod, by = c("X", "M1", "M2"))
```

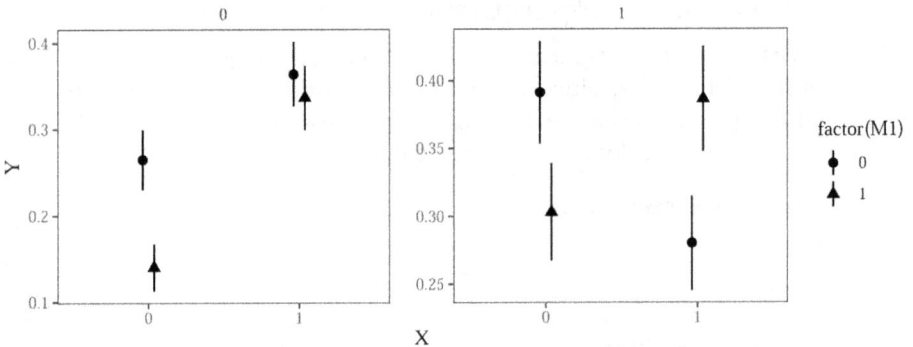

Figure 10.5: Average predicted outcomes for different combinations of X, M_1, and M_2.

Figure 10.5 shows the predicted probability that Y equals 1, for different combinations of X, M_1, and M_2. The x-axis represents the value of X. The shape of points represents the value of M_1, with triangles indicating a value of $M_1 = 1$. The left facet shows average predictions where $M_2 = 0$, and the right facet shows estimates for the subset of data where $M_2 = 1$.

In the left facet, we see that the average predicted probability that Y equals 1 is equal to about 26.5% when $X = 0$, $M_1 = 0$, and $M_2 = 0$. We can extract this numeric result by calling `avg_predictions()` on the relevant subset of data.

```
avg_predictions(mod, newdata = datagrid(
  X = 0, M1 = 0, M2 = 0
))
```

Estimate	Std. Error	z	Pr(>|z|)	2.5 %	97.5 %
0.265	0.0177	15	<0.001	0.23	0.299

Similarly, we can use the by argument to compute average predicted probabilities for every combination of the predictors (results omitted for brevity).

```
avg_predictions(mod, by = c("X", "M1", "M2"))
```

Does X affect Y?

As before, we can estimate the average change in Y associated with a change from 0 to 1 in the X variable using the avg_comparisons() function.

```
avg_comparisons(mod, variables = "X")
```

Estimate	Std. Error	z	Pr(>|z|)	2.5 %	97.5 %
0.0657	0.0129	5.1	<0.001	0.0405	0.091

This suggests that, on average, moving from the control (0) to the treatment (1) group is associated with an increase of 6.6 percentage points in the probability that Y equals 1. The p value is small, which implies that we can reject the null hypothesis that X has no effect on the predicted outcome.

Is the effect of X on Y moderated by M_1?

To check estimate if the effect of X varies depending on the value of moderator M_1, we call avg_comparisons() with the by argument.

```
avg_comparisons(mod, variables = "X", by = "M1")
```

M1	Estimate	Std. Error	z	Pr(>|z|)	2.5 %	97.5 %
0	−0.00703	0.0185	−0.38	0.704	−0.0433	0.0292
1	0.14026	0.0179	7.83	<0.001	0.1052	0.1754

This shows that the expected change in Y associated with a change in X differs based on the value of M_1: −0.0070 vs. 0.1403. Using the hypothesis argument, we can confirm that the difference between these two estimated effect sizes is statistically significant. In other words, we can reject the null hypothesis that the effect of X is the same for all values of M_1.

```
avg_comparisons(mod,
  variables = "X",
  by = "M1",
  hypothesis = "b2 - b1 = 0")
```

Hypothesis	Estimate	Std. Error	z	Pr(>|z|)	2.5 %	97.5 %
b2-b1=0	0.147	0.0258	5.72	<0.001	0.0968	0.198

Does the moderating effect of M_1 depend on M_2?

The last question that we pose is more complex. Above, we established two patterns.

1. On average, the value of X affects the predicted value of Y.
2. On average, the value of M_1 modifies the strength of association between X and Y.

Now, we ask if M_2 changes the way in which M_1 moderates the effect of X on Y. The difference is subtle but important: we are asking if the moderation effect of M_1 is itself moderated by M_2.

The following code computes the average difference in predicted Y associated with a change in X, for every combination of moderators M_1 and M_2. Each row represents the average effect of X at different points in the sample space.

```
avg_comparisons(mod,
    variables = "X",
    by = c("M2", "M1"))
```

M2	M1	Estimate	Std. Error	z	Pr(>|z|)	2.5 %	97.5 %
0	0	0.0992	0.0261	3.80	< 0.001	0.0481	0.1504
0	1	0.1967	0.0236	8.35	< 0.001	0.1505	0.2429
1	0	−0.1111	0.0262	−4.24	< 0.001	−0.1625	−0.0597
1	1	0.0834	0.0270	3.09	0.00203	0.0304	0.1363

When we hold the moderators fixed at $M_1 = 0$ and $M_2 = 0$, changing the value of X from 0 to 1 changes the average predicted probability of $Y = 1$ by 9.9 percentage points. When we hold the moderators fixed at $M_1 = 0$ and

$M_2 = 1$, changing the value of X from 0 to 1 changes the average predicted probability of $Y = 1$ by -11.1 percentage points.

Now, imagine we hold M_2 constant at 0. We can determine if the effect of X is moderated by M_1 by using the **hypothesis** argument to compare estimates in rows 1 and 2. This shows that the estimated effect size of X is larger when $M_1 = 1$ than when $M_1 = 0$, holding $M_2 = 0$.

```
avg_comparisons(mod,
    hypothesis = "b2 - b1 = 0",
    variables = "X",
    by = c("M2", "M1"))
```

Hypothesis	Estimate	Std. Error	z	Pr(>|z|)	2.5 %	97.5 %
b2-b1=0	0.0975	0.0351	2.77	0.00555	0.0286	0.166

Similarly, imagine that we hold M_2 constant at 1. We can determine if the effect of X is moderated by M_1 by comparing estimates in rows 3 and 4. We use the syntax introduced in Chapter 4, where **b3** represents the 3rd estimate and **b4** the fourth.

```
avg_comparisons(mod,
    hypothesis = "b4 - b3 = 0",
    variables = "X",
    by = c("M2", "M1"))
```

Hypothesis	Estimate	Std. Error	z	Pr(>|z|)	2.5 %	97.5 %
b4-b3=0	0.194	0.0377	5.16	<0.001	0.121	0.268

The hypothesis test above shows that we can reject the null hypothesis that the 4th estimate is equal to the 3rd (i.e., that the difference between them is zero). Therefore, we can conclude that the effect size of X is larger when $M_1 = 1$ than when $M_1 = 0$, holding M_2 at 1.

So far, we have assessed the extent to which M_1 acts as a moderator, holding M_2 at different values. To answer the question of whether M_2 moderates the moderation effect of M_1, we can specify the **hypothesis** as a difference in differences:

```
avg_comparisons(mod,
    hypothesis = "(b2 - b1) - (b4 - b3) = 0",
    variables = "X",
    by = c("M2", "M1"))
```

Hypothesis	Estimate	Std. Error	z	Pr(>|z|)	2.5 %	97.5 %
(b2-b1)-(b4-b3)=0	−0.097	0.0515	−1.88	0.0597	−0.198	0.00396

This suggests that M_2 *may* have a second-order moderation effect, but we cannot completely rule out the null hypothesis because the p value does not cross conventional thresholds of statistical significance (p=0.0597). In other words, we cannot quite reject the null hypothesis that M_2 has no moderation effect on the moderation effect of M_1.

10.2   Polynomial regression

Polynomial regression is an extension of the standard additive regression framework, that can be used to model the relationship between a dependent variable Y and an independent variable X as an nth-degree polynomial. While the model specification remains linear in the coefficients, it is polynomial in the value of X. This type of regression is useful when the data shows a non-linear relationship that a straight line cannot adequately capture.

The general form of a polynomial regression model is

$$Y = \beta_0 + \beta_1 X + \beta_2 X^2 + \beta_3 X^3 + \cdots + \beta_n X^n + \varepsilon,$$

where Y is the dependent variable, X is the independent variable, $\beta_0, \beta_1, \beta_2, \ldots, \beta_n$ are the coefficients to be estimated, n is the degree of the polynomial, and ε represents the error term. For instance, a second-degree (quadratic) polynomial regression equation can be written as:

$$Y = \beta_0 + \beta_1 X + \beta_2 X^2 + \varepsilon$$

A polynomial regression can be seen as a special case of the multiplicative interactions discussed in Section 10.1, where predictors are simply interacted with themselves. As before, they can be treated as a simple linear regression problem by constructing new variables $Z_1 = X$, $Z_2 = X^2$, etc. The model then becomes $Y = \beta_0 + \beta_1 \cdot Z_1 + \beta_2 \cdot Z_2 + \varepsilon$, which can be estimated using standard methods like ordinary least squares.

Polynomial regression offers several key advantages. It is flexible and can fit a wide range of curves, simply by adjusting the degree of the polynomial. As a result, polynomial regression can reveal underlying patterns in the data that are not immediately apparent with simpler models.

This approach also has notable disadvantages. One significant issue is its potential for overfitting, especially when the degree is high. Moreover, polynomial regression can suffer from unreliable extrapolation, where predictions made outside the range of the observed sample can become erratic and unrealistic. Consequently, while polynomial regression can be powerful, careful consideration must be given to the degree of the polynomial to balance fit and generalization effectively.

Polynomial regression can be viewed simply as a model specification with several variables interacted with themselves. As such, it can be interpreted using exactly the same tools discussed in the earlier part of this chapter. To illustrate, we consider two simple data generating processes adapted from Hainmueller et al. (2019). The first is

$$Y = 2.5 - X^2 + \varepsilon, \qquad \text{where } \varepsilon \sim N(0,1) \text{ and } X \sim U(-3,3)$$

If we fit a linear model with only X as predictor, the line of best fit will not be a good representation of these data. However, a cubic polynomial regression can easily detect the curvilinear relationship between X and Y. In R and Python, we can use similar syntax to specify polynomials directly in the model formula.[9] Then, we call `plot_predictions()` to visualize the predictions of the model.

```
library(patchwork)
dat = get_dataset("polynomial_01")

mod_linear = lm(Y ~ X, data = dat)
p1 = plot_predictions(mod_linear, condition = "X", points = .05) +
  ggtitle("Linear")

mod_cubic = lm(Y ~ X + I(X^2) + I(X^3), data = dat)
p2 = plot_predictions(mod_cubic, condition = "X", points = .05) +
  ggtitle("Cubic")

p1 + p2
```

---

[9]For `marginaleffects` to work properly in this context, it is important to specify the polynomials in the model-fitting formula. Users should *not* hard-code the values by creating new variables in the dataset before fitting the model.

Linear

Cubic

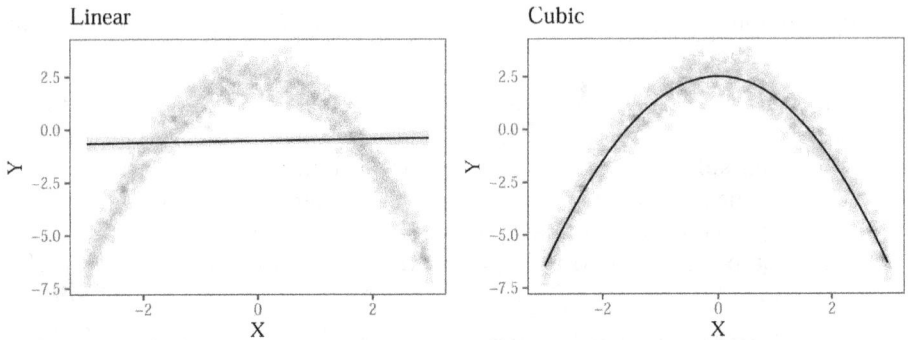

Figure 10.6: Modeling a curvilinear relationship with linear or polynomial regression.

On the left-hand side, we see the predictions of a linear model. The prediction line runs straight through the cloud of observed data points, and fails to capture the curvilinear relationship between X and Y. On the right-hand side, the prediction line drawn based on cubic terms does a very good job of capturing the curve in the data generating process.

To evaluate the strength of association between X and Y, we can compute the slope of the outcome equation with respect to X, at different values of X.

```
slopes(mod_cubic, variables = "X", newdata = datagrid(X = c(-2, 0, 2)))
```

X	Estimate	Std. Error	z	Pr(>|z|)	2.5 %	97.5 %
−2	3.98616	0.0351	113.596	<0.001	3.9174	4.0549
0	0.00478	0.0226	0.211	0.833	−0.0396	0.0491
2	−3.97639	0.0359	−110.705	<0.001	−4.0468	−3.9060

When X is negative, the slope is positive which indicates that an increase of X is associated with an increase in Y. When X is around 0, the slope is null, which indicates that the strength of association between X and Y is null (or very weak). When X is around 0, changing X by a small amount will have almost no effect on Y. When X is positive, the slope is negative. This indicates that increasing X should result in a decrease in Y.

Now, consider a slightly different data generating process, where a binary moderator M changes the nature of the relationship between X and Y.

$$Y = 2.5 - X^2 - 5 \cdot M + 2 \cdot M \cdot X^2 + \varepsilon \text{ where } \varepsilon \sim N(0,1) \text{ and } X \sim U(-3,3)$$

If we simply fit a cubic regression, without accounting for M, our predictions will be inaccurate. However, if we interact the moderator M with all polynomial terms (using parentheses as a shortcut for the distributive property), we can get an excellent fit for the curvilinear and differentiated relationship between X and Y.

```
dat = get_dataset("polynomial_02")

mod_cubic = lm(Y ~ X + I(X^2) + I(X^3), data = dat)
p1 = plot_predictions(mod_cubic, condition = "X",
  points = .05)

mod_cubic_interaction = lm(Y ~ M * (X + I(X^2) + I(X^3)), data = dat)
p2 = plot_predictions(mod_cubic_interaction, condition = c("X", "M"),
  points = .1)

p1 + p2
```

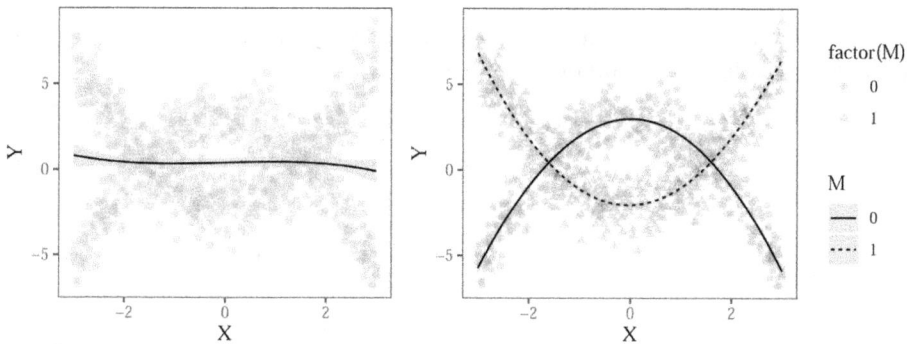

Figure 10.7: Predictions from two regression models fit to the same dataset.

As expected, the fit lines drawn by the model with cubic polynomials and an interaction do well in capturing the distinct patterns, in the two M subsets. And as before, we can estimate the slope of the outcome equation for different values of M and X:

```
s = slopes(mod_cubic_interaction,
  variables = "X",
  newdata = datagrid(M = c(0, 1), X = fivenum))
s
```

M	X	Estimate	Std. Error	z	Pr(>|z|)	2.5 %	97.5 %
0	−2.9974	5.7529	0.2588	22.228	<0.001	5.24566	6.2602
0	−1.4845	2.9009	0.0582	49.875	<0.001	2.78693	3.0149
0	−0.0586	0.1324	0.0667	1.986	0.047	0.00174	0.2631
0	1.4788	−2.9405	0.0576	−51.094	<0.001	−3.05325	−2.8277
0	2.9843	−6.0373	0.2542	−23.754	<0.001	−6.53549	−5.5392
1	−2.9974	−6.1017	0.2526	−24.156	<0.001	−6.59678	−5.6066
1	−1.4845	−2.9061	0.0563	−51.628	<0.001	−3.01646	−2.7958
1	−0.0586	−0.0461	0.0634	−0.727	0.467	−0.17026	0.0781
1	1.4788	2.8730	0.0595	48.284	<0.001	2.75638	2.9896
1	2.9843	5.5654	0.2596	21.437	<0.001	5.05657	6.0742

When $M = 0$ and $X \approx -3$, the estimated slope is 5.753, which means that the prediction curve is rising steeply. Indeed, this is what we see in the right panel of Figure 10.7, where the solid line rises quickly for low values of X. When $M = 0$ and $X \approx 3$, increasing X by a small amount will result in a substantial (and statistically significant) decrease in Y.

11

Categorical and ordinal outcomes

This chapter shows how the framework and tools introduced in Parts I and II help us give meaning to estimates obtained by fitting a categorical or ordinal outcome model.

We say that the outcome variable of a regression model is categorical when it is discrete that is, it involves a finite number of categories. Categorical outcomes are very common in data analysis, for example, when one wishes to model the choice of a mode of transportation (e.g., car, bus, bike, walk). A popular way to approach such data is to fit a multinomial logit model.

We say that the outcome variable of a regression model is ordinal when it is discrete, involves a finite number of categories, and has a natural ordering. Ordinal outcomes are very common in data analysis, for example, when one wishes to model levels of satisfaction (e.g., very dissatisfied, dissatisfied, neutral, satisfied, very satisfied). A popular way to approach such data is to fit an ordered probit model, where the probability of each category j of the outcome variable is modeled as (Cameron and Trivedi, 2005)

$$P(Y = j) = F(\theta_j - \mathbf{X}\beta) - F(\theta_{j-1} - \mathbf{X}\beta)$$

where $\mathbf{X}$ represents the vector of predictors, β the vector of coefficients, θ_j and θ_{j-1} the threshold parameters defining the boundaries between categories j and $j-1$, and F the standard normal cumulative distribution function.

The distinguishing feature of models like the multinomial logit or ordered probit is that they estimate different parameters for each level of the outcome variable. This allows the analyst to estimate different quantities of interest— predictions, counterfactual comparisons, and slopes—for each level of the dependent variable. This chapter focuses on the ordered probit case, but the same workflow applies to a much wider class of models, including other multinomial and ordered approaches.

Let's consider the dataset on extramarital affairs, collected in 1969 by the magazine *Psychology Today*, and analyzed in Fair (1978). These data include 601 survey responses on demographic, marital, and personal characteristics of married Americans: gender, age, years married, children, religiousness, education, occupation, and self-rated happiness in marriage.

DOI: 10.1201/978100356033-11

The outcome in our analyses is **affairs**, a variable that records the self-reported frequency of extramarital sexual encounters in the past year: 0, 1, 2, 3, 4–10 or >10. The following code loads the data and plots the distribution of the outcome.

```
library(MASS)
library(ggplot2)
library(marginaleffects)
dat = get_dataset("affairs")
barplot(table(dat$affairs), ylab = "N", xlab = "Affairs")
```

To study the determinants of the number of extramarital affairs, we specify an ordered probit regression model with three predictors: a binary variable indicating whether a survey respondent has children, the number of years since they got married, and a binary indicator for gender. To fit the model, we use the **polr()** function in the **MASS** package.

```
mod = polr(
  affairs ~ children + yearsmarried + gender,
  method = "probit", data = dat, Hess = TRUE)
summary(mod)
```

```
Coefficients:
                Value Std. Error t value
childrenyes   0.17290     0.1506  1.1477
yearsmarried  0.03457     0.0116  2.9793
genderwoman  -0.08851     0.1075 -0.8234
```

```
Intercepts:
          Value   Std. Error t value
0|1       1.0553  0.1336      7.8970
1|2       1.2524  0.1357      9.2300
2|3       1.3647  0.1372      9.9471
3|4-10    1.5057  0.1394     10.8010
4-10|>10  1.9363  0.1495     12.9536
```

The coefficients in this model can be viewed as measuring changes in a latent variable, associated to changes in the predictors. Unfortunately, building intuition about this latent variable is complicated, and the parameter estimates do not have a straightforward interpretation. Instead of focusing on those estimates, we do what we have done throughout this book, and transform raw parameters into more intuitive quantities of interest: predictions, counterfactual comparisons, and slopes.

11.1 Predictions

The first step of our post-estimation workflow is to compute average predictions. In a categorical outcome model, predictions are expressed on a probability scale. Importantly, the model allows us to make one prediction for every level of the outcome variable.

Imagine that we are specifically interested in the expected number of affairs for a woman with children who has been married for 10 years. To compute this quantity, we use the `datagrid()` function to build a grid, and call the `predictions()` function.[1]

```
p = predictions(mod, newdata = datagrid(
  children = "yes",
  yearsmarried = 10,
  gender = "woman"))
p
```

Group	children	years married	gender	Estimate	Std. Error	z	Pr(>|z|)	2.5 %	97.5 %
0	yes	10	woman	0.7341	0.02755	26.65	<0.001	0.6801	0.7881
1	yes	10	woman	0.0605	0.01041	5.81	<0.001	0.0401	0.0809
2	yes	10	woman	0.0304	0.00746	4.08	<0.001	0.0158	0.0451
3	yes	10	woman	0.0339	0.00795	4.27	<0.001	0.0184	0.0495
4–10	yes	10	woman	0.0750	0.01262	5.94	<0.001	0.0503	0.0998
>10	yes	10	woman	0.0660	0.01310	5.04	<0.001	0.0403	0.0917

This fitted model predicts that there is a 73% probability that this woman will have no extra-marital affair, and a 6% chance that she will have one.

Instead of reporting predicted probabilities for individual cases, we can compute them for all individuals in the dataset, and then average them. As in previous chapters, calling the `avg_predictions()` returns marginal predictions.

```
p = avg_predictions(mod)
p
```

---

[1]Chapter 5

Group	Estimate	Std. Error	z	Pr(>|z|)	2.5 %	97.5 %
0	0.7497	0.01739	43.12	<0.001	0.7157	0.7838
1	0.0567	0.00944	6.01	<0.001	0.0382	0.0752
2	0.0285	0.00680	4.19	<0.001	0.0152	0.0418
3	0.0317	0.00715	4.44	<0.001	0.0177	0.0457
4–10	0.0702	0.01039	6.76	<0.001	0.0498	0.0906
>10	0.0631	0.00986	6.40	<0.001	0.0438	0.0824

Since we are working with a categorical outcome model, the `avg_predictions()` function automatically returns one average predicted probability per outcome level. The group identifiers are available in the `group` column of the output data frame.

```
colnames(p)
```

```
[1] "group"     "estimate"   "std.error" "statistic" "p.value"    "s.value"
[7] "conf.low"  "conf.high"  "df"
```

```
p$group
```

```
[1] 0   1   2   3   4-10 >10
Levels: 0 1 2 3 4-10 >10
```

Unsurprisingly, given Figure 11.1, the expected probability that `affairs=0` is much higher than that of any of the other outcome levels (75%). In contrast, the average predicted probability that survey respondents have over 10 extramarital affairs in a year is 6%.

As we saw in Chapter 5, it is easy to compute average predictions for different subgroups. For instance, if we want to know the expected probability of each

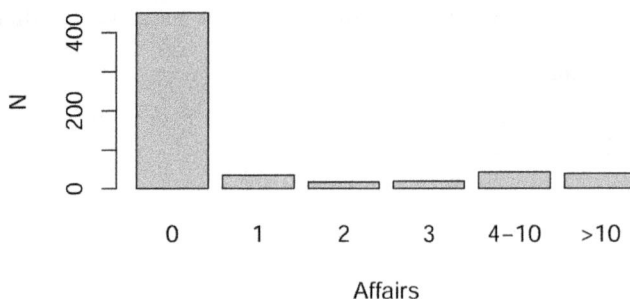

Figure 11.1: Distribution of the self-reported number of extra-marital affairs during 12 months preceding the survey (Fair, 1978).

outcome level, for survey respondents with and without children, we call the following code.

```
p = avg_predictions(mod, by = "children")
head(p, 2)
```

Group	children	Estimate	Std. Error	z	Pr(>|z|)	2.5 %	97.5 %
0	no	0.839	0.0275	30.5	<0.001	0.785	0.893
0	yes	0.714	0.0214	33.5	<0.001	0.672	0.756

On average, the predicted probability that a survey respondent without children has zero affairs is 84%. In contrast, the average predicted probability for a respondent with children is 71%.

We can also plot these figures with an analogous call, using the by argument to marginalize predictions across combinations of outcome group and children predictor.

```
plot_predictions(mod, by = c("group", "children")) +
  labs(x = "Number of affairs", y = "Average predicted probability")
```

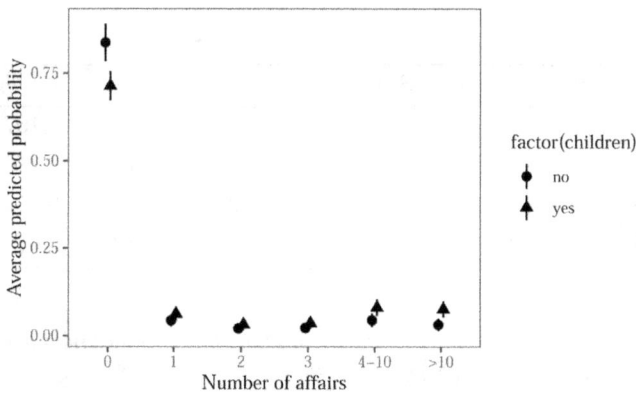

Figure 11.2: Average predicted probabilities for each outcome level, for the subpopulations with and without children.

Figure 11.2 shows again that the most likely survey response, by far, is zero. Moreover, the graph suggests that differences in average predicted probabilities between people with and without children are quite small. This impression will be confirmed in the next section, where we adopt an explicitly counterfactual approach.

In some cases, it can be useful to combine different outcome level categories. For example, the analyst may wish to know what the chances are that survey respondent will report at least one affair. To do this, we must compute the sum of all predicted probabilities for group values above 0.

One powerful strategy is to define a custom function to use in the hypothesis argument of avg_predictions(). As stated in the function manual,[2] custom hypothesis functions must accept the same data frame that we obtain when calling a function without the hypothesis argument, that is, a data frame with group, term, and estimate columns. The custom function can then process this input and return a new data frame with term and estimate columns.

For example, the custom function defined below accepts a data frame x, collapses the predicted probabilities by group categories, and returns a data frame of new estimates.

```
hyp = function(x) {
   x$term = ifelse(x$group == "0", "0", ">0")
   aggregate(estimate ~ term + children, FUN = sum, data = x)
}
p = avg_predictions(mod, by = "children", hypothesis = hyp)
p
```

Term	children	Estimate	Std. Error	z	Pr(>|z|)	2.5 %	97.5 %
>0	no	0.161	0.0275	5.87	<0.001	0.107	0.215
0	no	0.839	0.0275	30.53	<0.001	0.785	0.893
>0	yes	0.286	0.0214	13.38	<0.001	0.244	0.328
0	yes	0.714	0.0214	33.45	<0.001	0.672	0.756

On average, the predicted probability that a survey respondent without children will report at least one affair is 16%.

The custom function above applied a sum by subgroup using the aggregate() function, which has been available in base R since the 20th century. The same result can be obtained using the tidyverse idiom.

---

[2]In R, you can access the function manual by typing ?avg_predictions. In Python, it can be accessed by typing help(avg_predictions). The manual is also hosted at https://marginaleffects.com

```
library(tidyverse)
hyp = function(x) {
  x %>%
    mutate(term = if_else(group == "0", "0", ">0")) %>%
    summarize(estimate = sum(estimate), .by = c("term", "children"))
}
avg_predictions(mod, by = "children", hypothesis = hyp)
```

11.2 Counterfactual comparisons

After looking at predictions, it makes sense to adopt an explicitly counterfactual perspective, to quantity the strength of association between predictors and outcome. Our first objective is to answer this question: What would happen to the reported number of affairs if we increased the number of years married, while holding other predictors constant?

To answer this question, we use the `avg_comparisons()` function, and define the shift in focal predictor using the `variables` argument.[3]

```
avg_comparisons(mod, variables = list(yearsmarried = 5))
```

Group	Estimate	Std. Error	z	$Pr(>|z|)$	2.5 %	97.5 %
0	−0.05617	0.01940	-2.89	0.00379	-0.094199	-0.01814
1	0.00687	0.00236	2.91	0.00363	0.002241	0.01150
2	0.00427	0.00168	2.54	0.01094	0.000981	0.00756
3	0.00551	0.00215	2.56	0.01040	0.001296	0.00973
4–10	0.01590	0.00580	2.74	0.00607	0.004543	0.02726
>10	0.02362	0.00923	2.56	0.01053	0.005521	0.04171

```
cmp = avg_comparisons(mod, variables = list(yearsmarried = 5))
```

On average, increasing the `yearsmarried` variable by 5 reduces the predicted probability that `affairs` equals zero by 5.6 percentage points. On average, increasing `yearsmarried` by 5 increases the predicted probability that `affairs` is greater than 10 by 2.4 percentage points. These counterfactual comparisons are associated with small p values, so we can reject the null hypothesis that `yearsmarried` has no effect on the predicted number of `affairs`.

---

[3]Section 6.2.1

In contrast, if we estimate the counterfactual effect of gender, we find that differences in predicted probabilities are small and statistically insignificant. We cannot reject the null hypothesis that gender has no effect on affairs.

```
avg_comparisons(mod, variables = "gender")
```

Group	Estimate	Std. Error	z	$Pr(>|z|)$	2.5 %	97.5 %
0	0.02736	0.03324	0.823	0.411	−0.03779	0.09251
1	−0.00373	0.00458	−0.815	0.415	−0.01271	0.00524
2	−0.00225	0.00278	−0.808	0.419	−0.00769	0.00320
3	−0.00284	0.00350	−0.811	0.417	−0.00970	0.00402
4–10	−0.00786	0.00959	−0.819	0.413	−0.02666	0.01095
>10	−0.01068	0.01305	−0.818	0.413	−0.03626	0.01491

In conclusion, this chapter has demonstrated how categorical and ordinal outcome models, such as ordered probit, can be effectively used to interpret complex data. By transforming raw parameter estimates into intuitive quantities like predictions and counterfactual comparisons, analysts can gain meaningful insights into the data. This approach not only enhances the interpretability of the model results but also facilitates clear communication of findings to a broader audience.

12

Multilevel regression with poststratification

This chapter shows how `marginaleffects` can help us make sense of complex hierarchical data and draw inference from unrepresentative samples. To that end, we explore an empirical strategy called *multilevel regression with poststratification* (MRP).[1]

This MRP case study has three main objectives. First, it shows how to use `marginaleffects` to interpret estimates obtained by fitting multilevel regression models (aka, hierarchical, mixed effects, or random effects models). Second, it demonstrates that a consistent post-estimation workflow can be used to interpret the results of both frequentist and Bayesian models. Finally, this case study illustrates how to use poststratification to account for unrepresentative sampling.

We apply MRP to survey data collected in the United States. Our substantive goal is to estimate the level of support, in each American State, for budget cuts to police forces. This analysis follows Ornstein (2023), who compiled and studied data from the 2020 Cooperative Election Study (CES). Readers who are interested in a deeper dive into MRP, including a rich discussion of modeling choices and approaches, should refer to Ornstein's excellent tutorial.

In our analysis, the outcome of interest is `defund`, a binary variable equal to 1 if a respondent supports cuts to police budgets.[2] Predictors include individual-level covariates `gender`, `race`, `age`, and `education`. The dataset also records if each survey respondent has been a member of the `military`, and which `state` they live in.

The `ces_survey` dataset includes 3000 observations, randomly drawn from the full CES data.

---

[1]This case study only includes code for R. The features illustrated here are on the roadmap for `Python` but, at the time of writing, were not supported yet.

[2]The survey firm asked respondents if they supported or opposed this policy: "Decrease the number of police on the street by 10 percent, and increase funding for other public services."

```
library(marginaleffects)
survey = get_dataset("ces_survey")
head(survey)
```

```
# A data frame: 6 x 6
  state defund gender age    education    military
* <chr> <dbl>  <chr>  <fct>  <fct>            <dbl>
1 TX        0  Woman  70+    Some college         1
2 NJ        1  Woman  18-29  4 year               0
3 CO        0  Woman  60-69  4 year               1
4 PA        0  Man    50-59  4 year               1
5 PA        0  Woman  70+    High school          1
6 IA        1  Man    40-49  Some college         0
```

12.1   Multilevel models

Multilevel (or mixed effects) models are a popular strategy to study data that have a hierarchical, nested, or multilevel structure. Examples of nested data include survey respondents in different states; repeated measures made on the same subjects or on clustered observations; and students nested in classrooms, schools, districts, and states. A thorough introduction to multilevel modeling lies outside the scope of this chapter, but interested readers can refer to many great texts on the subject (Gelman and Hill, 2006; Finch et al., 2019; Hodges, 2021; Bürkner, 2024).

The parameters of a multilevel model can be divided into two types: "fixed effects" and "random effects." Fixed effects are parameters that are assumed to be constant across all groups in the population. Random effects, on the other hand, are parameters that are allowed to vary across groups. A major benefit of mixed effects models is that they allow variation in parameters between subsets of the data. For example, by specifying a "random intercept," the analyst can let the baseline level of support for **defund** vary across US states. By including a "random coefficient" for the **gender** variable, they can allow the strength of association between **gender** and **defund** to vary from state to state.

Importantly, the random parameters of a mixed effects model are not completely free to vary from group to group. Instead, the model typically imposes constraints on the shape of the distribution of parameters. In effect, this regularizes parameters, allowing estimates for groups with small sample sizes to be informed by estimates for groups with larger sample sizes, and pulling extreme observations closer to the middle of the distribution. This can reduce the risk of overfitting and stabilize estimates.

12.2 Frequentist

Several software packages in R allow us to fit mixed effects models from a frequentist perspective, such as lme4 (Bates et al., 2015) and glmmTMB (Brooks et al., 2017). When using one of these packages, we define models using the familiar formula syntax. The fixed components of our model are specified as we would any other predictor when fitting a linear or generalized linear model. Random effects components are specified using a special syntax with parentheses and a vertical bar.

y = x + z + (1 + z | group)

The formula above indicates that y is the outcome variable; the associations between predictors x and z, and the outcome y are captured by fixed effect parameters; 1 is a random intercept which allows the baseline level of y to vary across groups; z is a random coefficient, allowing the strength of association between z and y to vary across groups; and group is the grouping variable.

To model the probability of supporting cuts to police budgets, we use the glmmTMB package and fit a logistic regression model which includes a random intercept and random gender coefficient by state.

```
library(glmmTMB)
library(marginaleffects)

mod = glmmTMB(
    defund ~ age + education + military + gender + (1 + gender | state),
    family = binomial,
    data = survey)
summary(mod)

 Family: binomial  ( logit )
Formula:          defund ~ age + education + military + gender + (1 + gender |
    state)
Data: survey

     AIC      BIC   logLik -2*log(L)  df.resid
  3822.7   3918.8  -1895.3    3790.7      2984

Random effects:

Conditional model:
 Groups Name        Variance Std.Dev. Corr
 state  (Intercept) 0.004169 0.06457
        genderWoman 0.001333 0.03650  1.00
Number of obs: 3000, groups:  state, 48
```

```
Conditional model:
                      Estimate Std. Error z value Pr(>|z|)
(Intercept)          -0.2577928  0.2324920  -1.109  0.26751
age30-39              0.0192253  0.1256230   0.153  0.87837
age40-49             -0.7821669  0.1342192  -5.828 5.63e-09 ***
age50-59             -0.9662776  0.1278805  -7.556 4.15e-14 ***
age60-69             -1.1645794  0.1325723  -8.784  < 2e-16 ***
age70+               -1.3740248  0.1527391  -8.996  < 2e-16 ***
educationHigh school  0.0890342  0.2286540   0.389  0.69699
educationSome college 0.4157259  0.2297095   1.810  0.07033 .
education2 year       0.1891399  0.2478493   0.763  0.44539
education4 year       0.6204406  0.2294903   2.704  0.00686 **
educationPost grad    0.9882507  0.2405607   4.108 3.99e-05 ***
military              0.0003398  0.0829750   0.004  0.99673
genderWoman           0.1680639  0.0814649   2.063  0.03911 *
---
Signif. codes:  0 '***' 0.001 '**' 0.01 '*' 0.05 '.' 0.1 ' ' 1
```

The signs of these estimated coefficients are interesting, but their magnitudes are somewhat difficult to interpret. As in previous chapters, we can use marginaleffects to make sense of those estimates.

To start, let's use the predictions() function to compute the predicted probability of supporting cuts to police budgets for two hypothetical individuals: one from California and one from Alabama.[3]

```
p = predictions(mod, newdata = datagrid(
  state = c("CA", "AL"),
  gender = "Man",
  military = 0,
  education = "4 year",
  age = "50-59"
))
p
```

state	gender	military	education	age	Estimate	Std. Error	z	Pr(>|z|)	2.5 %	97.5 %
CA	Man	0	4 year	50–59	0.372	0.0308	12.1	<0.001	0.312	0.432
AL	Man	0	4 year	50–59	0.353	0.0301	11.7	<0.001	0.294	0.412

Our model predicts that a survey respondent with these demographic characteristics has a 37.2% chance of supporting funding cuts if they live in California, but a 35.3% chance if they live in Alabama.

To assess the strength of association between age and defund, we can use the avg_comparisons() function.

---

[3]Note that the uncertainty estimates reported by marginaleffects for frequentist mixed effects models only take into account the variability in fixed effects parameters, and not in random effects parameters. Bayesian modeling allows for more options in the quantification of uncertainty.

```
avg_comparisons(mod, variables = "age")
```

Contrast	Estimate	Std. Error	z	Pr(>|z|)	2.5 %	97.5 %
30–39 − 18–29	0.00462	0.0302	0.153	0.878	−0.0545	0.0638
40–49 − 18–29	−0.18801	0.0313	−6.002	<0.001	−0.2494	−0.1266
50–59 − 18–29	−0.22917	0.0292	−7.853	<0.001	−0.2864	−0.1720
60–69 − 18–29	−0.27087	0.0291	−9.295	<0.001	−0.3280	−0.2138
70+ − 18–29	−0.31145	0.0313	−9.938	<0.001	−0.3729	−0.2500

The **age** of survey respondents is strongly related to their propensity to answer "yes" on the **defund** question. On average, our model predicts that moving from the younger age bracket (18–29) to the older one (70+) is associated with a reduction of 31 percentage points in the probability of supporting cuts to police staff.

This brief example shows that we can apply the workflow described in earlier chapters to interpret the results of mixed effects models. In the next section, we analyze a similar model from a Bayesian perspective.

12.3 Bayesian

Bayesian regression analysis has a long history in statistics. Recent developments in computing, algorithm design, and software development have dramatically lowered the barriers to entry; they have made it much easier for analysts to estimate Bayesian mixed effects models.[4]

A typical Bayesian analysis involves several (often iterative) steps, including model formulation, prior specification, model refinement, estimation, and interpretation. Together, these steps form a "Bayesian workflow" that allows analysts to go from data to insight (Gelman et al., 2020).

This section illustrates how to use the **brms** package for R to fit Bayesian mixed effects models, and how the **marginaleffects** package facilitates two important steps of the Bayesian workflow: prior predictive checks and posterior summaries.

The model that we consider here has the same basic structure as the frequentist model from the previous section.

---

[4]A comprehensive treatment of Bayesian modeling would obviously require more space than this short chapter affords. Interested readers can refer to many recent texts, such as Gelman et al. (2013), McElreath (2020), or Bürkner (2024).

```
defund ~ age + education + military + gender + (1 + gender | state)
```

The `age`, `education`, and `military` variables are associated to fixed effect parameters, while the `gender` and intercept parameters are allowed to vary from state to state.

12.3.1 Prior predictive checks

One important difference between Bayesian and frequentist analysts, is that the former must explicitly specify priors over the parameters of their models. Priors encode the knowledge, beliefs, or information that the analyst held about the parameters of interest, before looking at the data. Some priors can have an important impact on the results, so it is crucial to choose them carefully.

Using the `prior` function from the `brms` package for R, we can easily specify different priors. For example, if there is a lot of uncertainty about the value of the parameters in our model, we can set "vague" or "diffuse" priors, say from a normal distribution with a large variance.

```
library(brms)
library(ggdist)
library(ggplot2)

priors_vague = c(
  prior(normal(0, 1e6), class = "b"),
  prior(normal(0, 1e6), class = "Intercept")
)
```

In contrast, if the analyst is confident that the parameters should be closer to zero, they can choose a more "informative" or "narrow" prior.

```
priors_informative = c(
  prior(normal(0, 0.2), class = "b"),
  prior(normal(0, 0.2), class = "Intercept")
)
```

For many analysts, choosing priors is a difficult exercise. Part of this difficulty is linked to the fact that we are often required to specify priors for parameters that are expressed on an unintuitive scale.[5] What do the coefficients of a mixed-effect logit model mean, and does it make sense to think that they are distributed normally with a variance of 0.2?

To help answer this question, we can conduct prior predictive checks, that is, we can simulate quantities of interest from the pure model and priors, without considering any data at all. This allows us to check if different priors

---

[5] Chapter 3

make sense, by looking at simulated quantities of interest with more intuitive interpretations than raw coefficients. We can then refine the model purely based on our prior knowledge, before we let our views be influenced by the observed data.

To conduct a prior predictive check in R, we begin by using the brm() function and its formula syntax to define the model structure, with fixed and random parameters. Then, we set family=bernoulli to specify a logistic regression model. The prior argument is used to specify the priors we want to use, either diffuse or informative. Finally, we instruct brms to ignore the actual data altogether, and to draw simulations only from priors, by setting the argument sample_prior="only".

```
model_vague = brm(
    defund ~ age + education + military + gender + (1 + gender | state),
    family = bernoulli,
    prior = priors_vague,
    sample_prior = "only",
    data = survey)

model_informative = brm(
    defund ~ age + education + military + gender + (1 + gender | state),
    family = bernoulli,
    prior = priors_informative,
    sample_prior = "only",
    data = survey)
```

The sample_prior argument is powerful. It allows us to draw simulated coefficients from the model and priors, without looking at the data. But one problem remains: the outputs of this function are still expressed on the same unintuitive scale as the model parameters.

```
fixef(model_vague)
```

	Estimate	Est.Error	Q2.5	Q97.5
Intercept	-1358.4595	1400823.4	-2781831	2719043
age30M39	19503.7905	996759.2	-1899753	1963344
age40M49	3523.6830	1003149.8	-1894097	1951398
age50M59	-13388.5194	995769.3	-1914897	2004686
age60M69	-10833.3448	1031207.5	-1970037	2031937
age70P	27095.2811	1026253.7	-1996471	2019135
educationHighschool	-23835.2111	986739.4	-1929888	1907825
educationSomecollege	-582.7351	1019188.2	-1989659	1996451
education2year	7462.3930	986440.6	-1936941	1938873
education4year	4361.3645	1000424.8	-1887777	1952272
educationPostgrad	-9154.9701	1006088.9	-1997184	1902682
military	-16838.0852	993453.9	-1965510	1932634
genderWoman	8475.6834	990286.1	-1913603	1911330

How can we know if the results printed above make sense? One simple approach is to post-process the prior-only **brms** model using `marginaleffects`. With this strategy, we can compute any of the quantities of interest introduced earlier in this book: predictions, counterfactual comparisons, and slopes.[6]

When using vague normal priors, the average predicted probability of supporting cuts to police budgets is centered at 0.5, with credible intervals that cover nearly the full unit interval.

```
p_vague = avg_predictions(model_vague, by = "gender")
p_vague
```

gender	Estimate	2.5 %	97.5 %
Man	0.491	0.0118	0.988
Woman	0.500	0.0232	0.984

Unsurprisingly, the intervals are much narrower when we use informative priors.

```
p_informative = avg_predictions(model_informative, by = "gender")
p_informative
```

gender	Estimate	2.5 %	97.5 %
Man	0.501	0.350	0.649
Woman	0.500	0.349	0.652

We can also plot the prior distribution of average predictions. First, we extract draws from both models using `get_draws()`. Then, we plot them using the **ggplot2** and **ggdist** packages.

```
draws = rbind(
  get_draws(p_vague, shape = "rvar"),
  get_draws(p_informative, shape = "rvar")
)
draws$Priors = c("Vague", "Vague", "Informative", "Informative")

ggplot(draws, aes(xdist = rvar, fill = Priors)) +
  stat_slabinterval(alpha = .5) +
  facet_wrap(~ gender) +
  scale_fill_grey() +
  labs(x = "Average predicted probability (prior-only)", y = "Density")
```

---

[6]Chapters 5, 6, 7

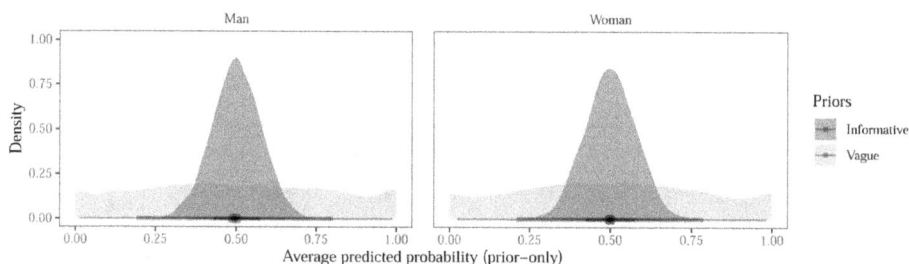

Figure 12.1: Average predictions with vague and informative priors (no data).

If the ultimate goal of our analysis is to compute an average treatment effect via G-computation,[7] we can conduct a prior predictive check directly on that quantity. Once again, the intervals are much wider when using vague priors.

```
avg_comparisons(model_vague, variables = "military")
```

Estimate	2.5 %	97.5 %
0	−0.523	0.506

```
avg_comparisons(model_informative, variables = "military")
```

Estimate	2.5 %	97.5 %
8.73e−05	−0.0557	0.0592

12.3.2 Posterior summaries

We now fit the model to data by dropping the `sample_prior` argument.

```
model = brm(
    defund ~ age + education + military + gender + (1 + gender | state),
    family = bernoulli,
    prior = priors_vague,
    data = survey)
```

Using this fitted model, we can consider the average predicted probability that a respondent answers "yes" on the `defund` question, for respondents with or without military experience. By default, `marginaleffects` functions report the mean of posterior draws, along with equal-tailed intervals.

---

[7]Chapters 6 and 8

```
p = avg_predictions(model, by = "military")
p
```

military	Estimate	2.5 %	97.5 %
0	0.447	0.424	0.472
1	0.364	0.341	0.387

On average, people who were never in the military have a 44.7% chance of supporting cuts to police budgets. In contrast, survey respondents with military backgrounds have, on average, an estimated probability of 36.4% of answering "yes."

As in Chapter 6, we can use the `avg_comparisons()` function to assess the strength of association between some covariate, like `age`, and the outcome `defund`.

```
cmp = avg_comparisons(model, variables = list(age = c("18-29", "70+")))
cmp
```

Estimate	2.5 %	97.5 %
−0.311	−0.37	−0.249

On average, holding all other variables constant at their observed values, moving from the 18–29 to the 70+ age categories would be associated with a decrease of −31 percentage points in the probability of supporting funding cuts.

Recall that our model specification allows the parameter associated to `gender` to vary from state to state. To see if this is the case, we call the `comparisons()` function with the `newdata` argument. This computes risk differences for one individual per state, whose characteristics are set to the average or mode of the predictors.

```
cmp = comparisons(model,
    variables = "gender",
    newdata = datagrid(state = unique))
```

Then, we sort the data, convert the `state` variable to a factor to preserve that order, and call `ggplot` to display the results.

```
cmp = sort_by(cmp, ~estimate) |>
    transform(state = factor(state, levels = state))

ggplot(cmp,
       aes(x = state, y = estimate,
           ymin = conf.low, ymax = conf.high)) +
    geom_hline(yintercept = 0, linetype = 3) +
    geom_pointrange() +
    labs(x = "", y = "Average risk difference") +
    theme(axis.text.x = element_text(angle = 90, vjust = .5))
```

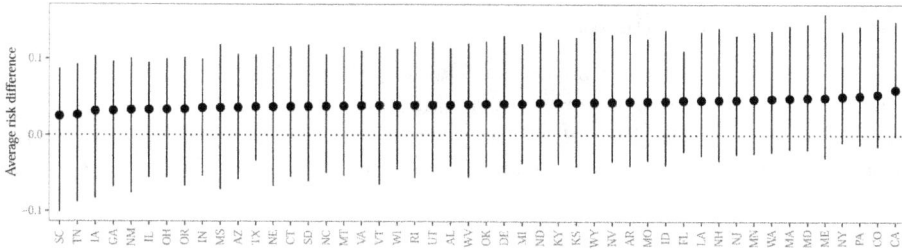

Figure 12.2: Average risk difference for a hypothetical individual with typical characteristics living in different states.

The results in Figure 12.2 show that the estimated risk difference is quite small, hovering around 4 percentage points. The plot also shows that the estimated association between **gender** and **defund** is quite stable across states.

When post-processing Bayesian results, it is often useful to directly manipulate draws from the posterior distribution. To extract these draws, we use the **get_draws()** function. This function can return several types of objects, including data frames in "wide" or "long" formats, matrices, or an **rvar** distribution object compatible with the **posterior** package. For instance, we can print the first 5 draws from the posterior distribution of average predictions with it.

```
draws = avg_predictions(model) |> get_draws(shape = "long")
draws[1:5, 1:2]
```

```
  drawid      draw
1      1 0.3964929
2      2 0.3974249
3      3 0.4046949
4      4 0.4067100
5      5 0.4161802
```

This allows us to compute various posterior summaries using standard R functions. The following code shows that about 70% of the posterior density of the average prediction lies above 0.4.

```
mean(draws$draw > .4)
```

```
[1] 0.69725
```

12.4 Poststratification

The sample that we used so far is relatively large: 3000 observations. But even if those observations were drawn randomly from the national population, they may still be insufficient to estimate state-level opinion. Indeed, some states are much more populous than others, and our dataset includes very few observations from certain areas of the country.

```
sort(table(survey$state))
```

```
ND  WY  SD  VT  MS  ID  MT  NE  NH  RI  DE  AR  KS  ME  WV  NM  OK  IA  NV  UT
 4   5   8   9  12  14  14  15  15  15  19  21  21  21  21  22  23  32  32  33
KY  CT  LA  MN  OR  AL  MD  MA  WA  CO  SC  WI  TN  IN  MO  AZ  NJ  GA  VA  NC
40  42  44  44  46  47  49  53  57  59  60  67  70  76  76  81  88  90  90  91
MI  IL  OH  PA  NY  FL  TX  CA
94 107 136 144 180 224 224 265
```

This kind of disparity occurs in many contexts, such as when one estimates quantities like

- Presidential voting intentions in swing states, based on a national survey.
- Psychological well-being of a population, using a web-based survey that oversamples young and highly educated people.
- Effect of a user interface change on website sales, based on an unbalanced sample of laptop or smartphone users.
- Vaccination rates using data from healthcare providers who are overrepresented in affluent neighborhoods.

To draw state-level inferences about defund, we will use the MRP approach, or multilevel regression with poststratification. MRP is implemented in four steps.

First, we estimate a mixed-effects model with random intercept by state. As Ornstein (2023) notes, the accuracy of MRP estimates depends on the predictive performance of this first-stage model, and may be affected by standard problems such as overfitting. Since we have already estimated such a model for previous examples, we simply use the same **brms** object as before: **model**.

Second, we must construct a poststratification frame, that is, a dataset that contains information about the prevalence of each possible predictor profile. Typically, the frame is constructed using a canonical source, such as the census. In our case, the **demographics** poststratification frame records the prevalence of socio-demographic characteristics for each state.[8]

```
demographics = get_dataset("ces_demographics")
head(demographics)
```

```
# A data frame: 6 x 6
  state age    education   gender military percent
* <chr> <fct>  <fct>       <chr>     <dbl>   <dbl>
1 AL    18-29  No HS       Man           0   0.318
2 AL    18-29  No HS       Man           1   0.212
3 AL    18-29  No HS       Woman         0   0.530
4 AL    18-29  No HS       Woman         1   0.318
5 AL    18-29  High school Man           0   1.80
6 AL    18-29  High school Man           1   0.636
```

This table shows, for example, that about 0.318% of the Alabama population are men between the ages of 18 and 29, with no high school degree, and no military experience. Note that the proportions of predictor profiles must sum to one for each state.

```
with(demographics, sum(percent[state == "AL"]))
```

```
[1] 100
```

The third step in MRP is to make predictions for each row of the poststratification frame. In other words, we use the **demographics** data frame in **newdata**, instead of using the empirical grid.

```
p = predictions(model, newdata = demographics)
```

Finally, we take a weighted average of these predictions by state, where weights are defined as the proportion of each demographic group in the state's population. To do this with marginaleffects, we modify the call above, specifying weights with the **wts** argument, and the aggregation unit with the **by** argument.

---

[8]Ideally, the **demographics** data frame would be "complete," in the sense that it would record the prevalence of every possible type of individual, in every state. Unfortunately, this is not always possible, when demographic information is not available for certain subsets of the population, for example in narrow groups, in less populous states.

```
p = avg_predictions(model,
  newdata = demographics,
  wts = "percent",
  by = "state")
head(p)
```

state	Estimate	2.5 %	97.5 %
AL	0.409	0.354	0.464
AR	0.399	0.346	0.474
AZ	0.381	0.334	0.430
CA	0.451	0.413	0.506
CO	0.436	0.397	0.505
CT	0.401	0.349	0.455

This command gives us one estimate per state. After adjusting for the demographic composition of the state of Alabama, our model estimates that the average probability of supporting cuts to police budgets is 40.9%, with a 95% credible interval of [35.4, 46.4].

To visualize these results, we can extract draws from the posterior distribution using the get_draws() function, and use the **ggdist** package to plot densities. Figure 12.3 shows that the support for cuts to police budgets varies across states, with highest estimates in California and lower estimates in Wyoming.

```
library(ggdist)
library(ggplot2)

p = p |>
  get_draws(shape = "rvar") |>
  sort_by(~ estimate) |>
  transform(state = factor(state, levels = state))

ggplot(p, aes(y = state, xdist = rvar)) +
  stat_slab(height = 2, color = "white") +
  labs(x = "Posterior density", y = NULL) +
  xlim(c(.3, .5))
```

In this chapter, we explored the use of multilevel regression with poststratification (MRP) to analyze hierarchical data and draw inferences from unrepresentative samples. We demonstrated how to fit mixed effects models using both frequentist and Bayesian approaches, and how to interpret these models using the **marginaleffects** package. The chapter highlighted the importance of prior predictive checks in Bayesian analysis and illustrated the process of poststratification to adjust for demographic composition.

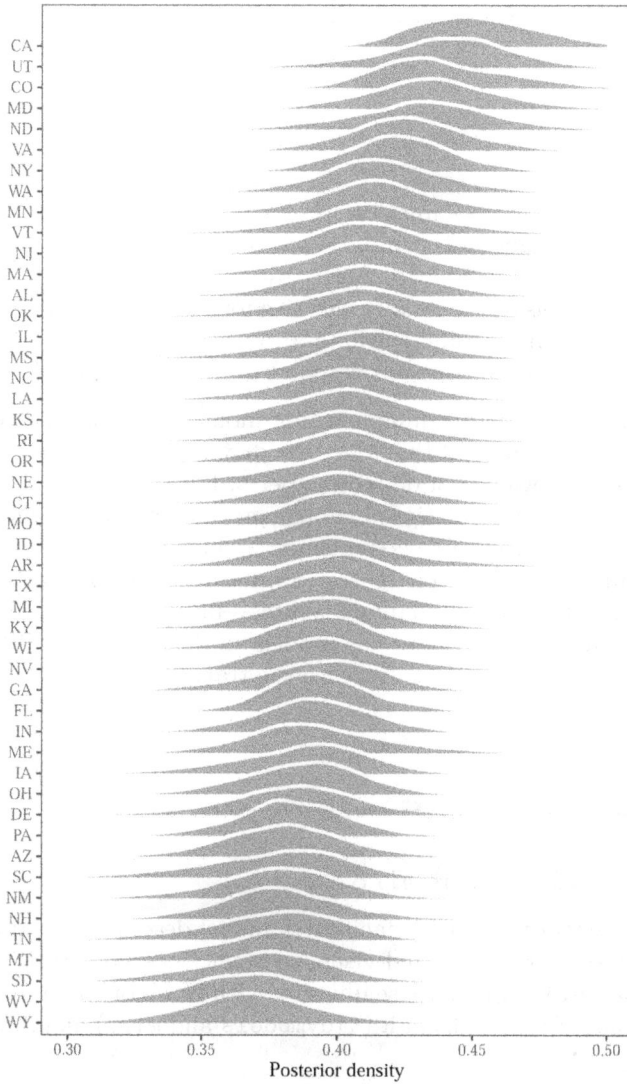

Figure 12.3: Estimated average probability of supporting cuts to police budgets, for each US state. Estimates obtained via Bayesian multilevel regression and poststratification.

13

Machine learning

The concepts and post-estimation tools introduced in earlier chapters—predictions, counterfactual comparisons, and slopes—are largely model-agnostic; they are applicable to both statistical and machine learning approaches. These tools are especially effective for model description, a task that is very important in machine learning applications, where analysts need to audit and understand how models respond to different inputs.[1]

Auditing and describing machine learning models is essential to ensure that predictions remain fair, and that they are driven by factors compatible with the substantive knowledge of domain experts. For instance, in credit scoring systems, evaluating how variations in applicant characteristics—such as income, employment status, or ethnicity—influence creditworthiness assessments helps detect and mitigate potential biases. Similarly, in hiring algorithms, how models weight different candidate attributes—like education level or years of experience—can help recruiters use models in decision-making. Audits and model description are crucial to improve the transparency and interpretability of data analyses.

13.1 tidymodels and mlr3

The `marginaleffects` package facilitates model description and auditing by allowing analysts to compute and visualize predictions, counterfactual comparisons, and slopes. It integrates seamlessly with some of the most prominent machine learning frameworks in R (`tidymodels` and `mlr3`) and Python (Scikit Learn).

`tidymodels` is a collection of packages in R designed for modeling and machine learning using `tidyverse` principles, offering a cohesive interface for data preprocessing, modeling, and validation (Kuhn and Wickham, 2020). `mlr3` is

---

[1] Section 2.1.1

a modern, object-oriented framework in R that provides a comprehensive suite
of tools for machine learning, including a wide array of algorithms, resampling
methods, and performance measures (Lang et al., 2019). By supporting both
tidymodels and mlr3, marginaleffects enables users to interpret a wide
variety of machine learning models . Scikit Learn is a powerful Python library
for machine learning that provides simple and efficient tools for data mining
and data analysis (Pedregosa et al., 2011).

A comprehensive introduction to machine learning in general, or to particular
frameworks, lies outside the scope of this book.[2] Instead, this chapter shows a
very simple example to demonstrate that the workflow built-up in previous
chapters applies in straightforward fashion to this new context.

Let us consider data on Airbnb rental properties in London, collected and
distributed by Békés and Kézdi (2021). This dataset includes information on
over 50,000 units, including features such as the unit type (single room or
entire unit), number of bedrooms, parking, or internet access. The primary
outcome of our analysis is the rental price of each unit.

To begin, we load the tidymodels and marginaleffects libraries, read the
data, and display the first rows and columns.

```
library(tidymodels)
library(marginaleffects)
set.seed(48103)
airbnb = get_dataset("airbnb")
airbnb[1:5, 1:6]
```

```
# A data frame: 5 x 6
  price bathrooms bedrooms  beds unit_type     `24-hour check-in`
* <int>     <dbl>    <int> <int> <chr>                      <int>
1    23         1        1     1 Private room                   0
2    50         1        1     1 Private room                   0
3    24         1        1     1 Private room                   0
4    50       1.5        1     1 Private room                   1
5    25         1        1     1 Private room                   0
```

The airbnb dataset includes 55 columns and 52717 rows. We split those rows
into a training set to fit the model, and a test set to make predictions and
evaluate the model's behavior.

```
airbnb_split = initial_split(airbnb)
train = training(airbnb_split)
test = testing(airbnb_split)
```

---

[2]See Kuhn and Silge (2022), James et al. (2023), Bischl et al. (2024)

The next block of code holds the core `tidymodels` fitting commands. The `boost_tree()` function specifies the model type. In this case, we use the XGBoost implementation of boosted trees, to predict a continuous outcome (i.e., "regression"). Users who prefer a different prediction algorithm could swap this line for `linear_reg()`, `rand_forest()`, `bart()`, etc. The `recipe()` function identifies the outcome variable (`price`), and initiates a data pre-processing "recipe," that is, a series of steps to transform the raw data into a suitable format for model fitting. `step_dummy()` is used to convert categorical predictors into dummy variables. Finally, the `workflow()` function combines the model and the pre-processing recipe, and the `fit()` function fits the model to the training data.

```
xgb = boost_tree(mode = "regression", engine = "xgboost")

mod = recipe(airbnb, price ~ .) |>
  step_dummy(all_nominal_predictors()) |>
  workflow(spec = xgb) |>
  fit(train)
```

13.2   Predictions

With the fitted model in hand, we now use the `predictions()` function to generate predictions in the test set. As usual, `predictions()` returns a simple data frame with the quantity of interest in the `estimate` column, and the original data in separate columns. Thus, we can easily check the quality of our predictions in the test set by plotting the predicted values of the outcome (`estimate`) against the actually observed values (`price`).

```
p = predictions(mod, newdata = test)

ggplot(p, aes(x = price, y = estimate)) +
  geom_point(alpha = .2) +
  geom_abline(linetype = 3) +
  labs(x = "Observed Price", y = "Predicted Price") +
  xlim(0, 500) + ylim(0, 500) +
  coord_equal()
```

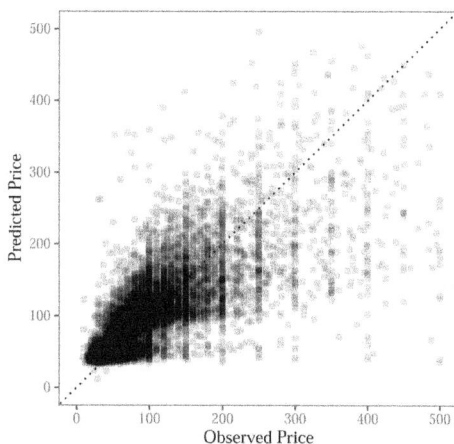

Figure 13.1: XGBoost predictions of rental prices against observed prices.

Every point in Figure 13.1 represents one rental unit in the test set. The x-axis shows the observed price for that unit and the y-axis shows the predicted price. Points on the diagonal are correctly predicted. There is considerable spread around that diagonal, which means our algorithm makes substantial prediction errors.

Most of the standard functions and arguments in **marginaleffects** are available. For instance, to compute the average predicted price of private rooms and entire homes in the test set, we call **avg_predictions()** with the by argument.[3]

```
avg_predictions(mod,
  by = "unit_type",
  newdata = test)
```

unit_type	Estimate
Entire home/apt	135.9
Private room	50.3

Unsurprisingly, our model expects that, on average, entire homes should be more expensive than private rooms.

---

[3]Note that when we apply a **marginaleffects** function to model fitted by **tidymodels** or **mlr3**, we do not obtain standard errors. This is because the parameters of machine learning estimates are not typically accompanied by a variance-covariance matrix, which implies that we cannot use the delta method. **tidymodels** has built-in support for some uncertainty quantification strategies like conformal prediction.

13.2.1   Partial dependence plot

A Partial Dependence Plot (PDP) is a strategy to visualize how predictions change with certain predictors. It computes predictions over a range of values for a predictor, averaging over other variables, to show how outcomes vary with changes in a feature. This is useful for understanding complex models.

The `plot_predictions()` function in the `marginaleffects` package simplifies the creation of these plots. The command below computes average predicted outcomes for each combination of `bedrooms` and `unit_type`, and plots the results.

```
plot_predictions(mod,
  by = c("bedrooms", "unit_type"),
  newdata = airbnb) +
  labs(x = "# Bedrooms", y = "Predicted Price", linetype = "")
```

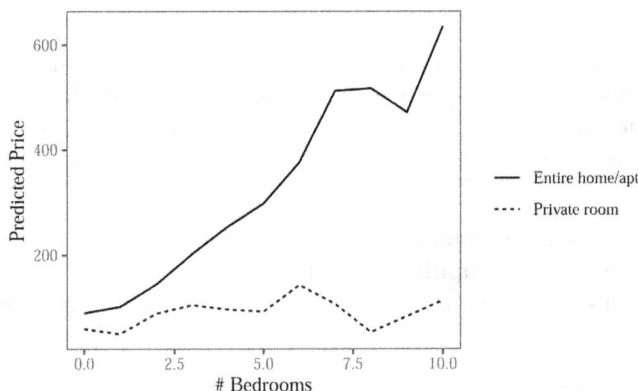

Figure 13.2: Relationship between predicted price, number of bedrooms, and type of rental unit in London.

The plot in Figure 13.2 makes sense substantively. On the one hand, the price of a single private room does not really change as we increase the total number of bedrooms in the unit. On the other, the price of renting an entire unit does increase with the number of bedrooms.

In some contexts, analysts prefer to draw partial dependence plots based on a counterfactual grid.[4] The idea, here, is to duplicate the entire dataset once for every combination of values of the focal variables. Then, we make predictions on that counterfactual grid, and take an average of predictions. This counterfactual approach ensures that the distribution of marginalized covariates is identical for every combination of predictors. The resulting plots can be interpreted as illustrating "all else equal" predictions.

---

[4]Section 5.2.5

To draw this kind of partial dependence plot, we first build a counterfactual grid. If the data is very large, duplicating it several times to create counterfactual versions can require a lot of memory. To circumvent this problem, we build the grid based on random subset of 10,000 rows from the data set.

```
set.seed(48103)
airbnb_subset = airbnb[sample(1:nrow(airbnb), 10000), ]

grid = datagrid(
  bedrooms = unique,
  unit_type = unique,
  newdata = airbnb_subset,
  grid_type = "counterfactual")

plot_predictions(mod,
  newdata = grid,
  by = c("bedrooms", "unit_type")) +
  labs(x = "# Bedrooms", y = "Predicted Price", linetype = "")
```

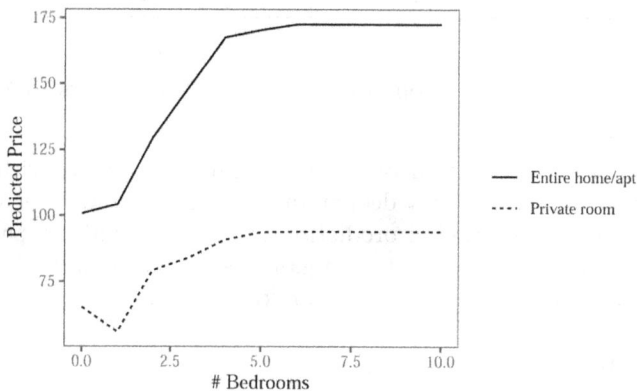

13.3   Counterfactual comparisons

As in Chapter 6, we can use the avg_comparisons() function to answer counterfactual queries such as: On average, how does the predicted price change when we increase the number of bedrooms by 2, holding all other variables constant?

```
avg_comparisons(mod,
  variables = list(bedrooms = 2),
  newdata = airbnb)
```

Estimate
27.3

Our model predicts that the price of a unit with two extra bedrooms will be £27 higher. Furthermore, we may inquire about the combined effect of increasing the number of bedrooms by one, and of transitioning from an apartment without wireless internet access to one with such access. For this, we use the `cross` argument.

```
avg_comparisons(mod,
    variables = c("bedrooms", "Wireless Internet"),
    cross = TRUE,
    newdata = airbnb)
```

C: bedrooms	C: Wireless Internet	Estimate
+1	1 − 0	14.6

On average, adding one bedroom and wireless internet access to a rental unit increases the expected price by 15.

In conclusion, the integration of machine learning models with tools like **marginaleffects** allows for a deeper understanding and interpretation of complex models. By leveraging predictions, counterfactual comparisons, and partial dependence plots, analysts can gain insights into model behavior and ensure that predictions align with domain knowledge. This approach not only enhances model transparency but also aids in making informed decisions based on model outputs.

14

Uncertainty

This chapter introduces four approaches to quantify uncertainty around quantities of interest: the delta method, bootstrap, simulation-based inference, and conformal prediction. We discuss the intuition behind each of these strategies, and reinforce that intuition by implementing simple versions "by hand." The code examples are presented for pedagogical purposes; more robust and convenient implementations are supplied by the `marginaleffects` package. Readers who are not familiar with the basics of multivariable calculus may want to skip this chapter.

14.1   Delta method

The delta method is a statistical technique used to approximate the variance of a function of random variables. For example, data analysts often want to compare regression coefficients or predicted probabilities by subtracting them from one another. Such comparisons can be expressed as functions of a model's parameters, but the variances (and standard errors) of transformed quantities can be difficult to derive analytically. The delta method is a convenient way to approximate those variances. It allows us to quantify uncertainty around most of the quantities of interest discussed in this book.

The delta method is useful, versatile, and fast. Yet, the technique also has important limitations: it relies on a coarse linear approximation, its desirable properties depend on the asymptotic normality of the estimator,[1] and it is only valid when a function is continuously differentiable in a neighborhood of its parameters. When analysts do not believe that these conditions hold, they can turn to the bootstrap or simulation-based inference.[2]

The theoretical properties of the delta method are discussed and proved in many textbooks (Wasserman, 2004; Hansen, 2022a; Berger and Casella, 2024).

---

[1] The delta method relies on the observation that if the distribution of a parameter is asymptotically normal, then the limiting distribution of a smooth function of this parameter is also normal (Wasserman, 2004).

[2] Sections 14.2 and 14.3

DOI: 10.1201/9781003560333-14

Instead of demonstrating them anew, the rest of this section provides a hands-on tutorial, with a focus on computation. The goal is to reinforce intuitions by illustrating how to calculate standard errors manually in two cases: univariate and multivariate. Then, we briefly consider the use of robust or clustered standard errors.

14.1.1 Univariate delta method

In this section, we use the delta method to compute the standard error of a function with one parameter: the natural logarithm of a single regression coefficient. Consider a setting with three measured variables: Y, X, and Z. With these variables in hand, the analyst estimates a linear regression model.

$$Y = \beta_1 + \beta_2 X + \beta_3 Z + \varepsilon \tag{14.1}$$

Let $\hat{\beta}_2$ be an estimator of the true parameter β_2. We know that the variance of $\hat{\beta}_2$ is $\text{Var}(\hat{\beta}_2)$, but we are not interested in this particular statistic. Rather, our goal is to compute the variance of a function of the estimator: $\text{Var}\left[\log(\hat{\beta}_2)\right]$.

To get there, we apply a Taylor series expansion. This mathematical tool allows us to rewrite smooth functions like $\log(\hat{\beta}_2)$ as infinite sums of derivatives and polynomials. This is a useful way to simplify and approximate complex functions.

The Taylor series of a function $g(a)$ is always expressed by reference to a particular point α, where we wish to study the behavior of g. Generically, the approximation can be written as

$$g(a) = g(\alpha) + g'(\alpha)(a - \alpha) + \frac{g''(\alpha)}{2!}(a - \alpha)^2 + \cdots + \frac{g^{(n)}(\alpha)}{n!}(a - \alpha)^n + \cdots, \tag{14.2}$$

where $g^{(n)}(\alpha)$ denotes the nth derivative of $g(a)$, evaluated at $a = \alpha$.

Equation 14.2 is an infinite series, so it is not practical (or possible) to compute all its terms. In real-world applications, we thus truncate the series by dropping higher-order terms. For instance, we can construct a first-order Taylor series by keeping only the first two terms on the right-hand side of Equation 14.2.

We can now approximate the function that interests us, $\log(\hat{\beta}_2)$, as a first-order Taylor series, about the true value of the parameter β_2. Recall that the derivative of $\log(\beta_2)$ is $1/\beta_2$, and substitute appropriate values into the first two terms of Equation 14.2.

$$g(a) \approx g(\alpha) + g'(\alpha)(a - \alpha)$$

$$\log(\hat{\beta}_2) \approx \log(\beta_2) + \frac{1}{\beta_2}(\hat{\beta}_2 - \beta_2)$$

Now, insert the approximation in the variance operator.

$$\text{Var}\left[\log(\hat{\beta}_2)\right] \approx \text{Var}\left[\log(\beta_2) + \frac{1}{\beta_2}(\hat{\beta}_2 - \beta_2)\right] \tag{14.3}$$

This linearized approximation of the variance is useful, because it allows further simplification. First, we know that the variance of a constant like β_2 is zero, which means that some of the terms in Equation 14.3 can drop out. Second, we can apply the scaling property of variances[3] to rewrite the expression as

$$\text{Var}\left[\log(\hat{\beta}_2)\right] \approx \frac{1}{\beta_2^2}\text{Var}\left[\hat{\beta}_2\right] \tag{14.4}$$

Equation 14.4 gives us a formula to approximate the variance of $\log(\hat{\beta}_2)$. This formula is expressed in terms of $\text{Var}(\hat{\beta}_2)$, that is, the square of the estimated standard error. Equation 14.4 also includes the expression $1/\beta_2$. Since we do not know the true value of that term, we fall back to using $1/\hat{\beta}_2$ as a plug-in estimate.[4]

To show how this formula can be applied in practice, let's simulate data that conform to Equation 14.1, and use that data to fit a linear regression model.

```
set.seed(48103)
X = rnorm(100)
Z = rnorm(100)
Y = 1 * X + 0.5 * Z + rnorm(100)
dat = data.frame(Y, X, Z)
mod = lm(Y ~ X + Z, data = dat)
summary(mod)
```

	Estimate	Std. Error	t value	Pr(>|t|)
(Intercept)	-0.05033	0.09740	-0.517	0.6065
X	1.12963	0.09929	11.377	<0.001
Z	0.28486	0.11013	2.587	0.0112

The true β_2 is 1, but the estimated $\hat{\beta}_2$ is

```
b2 = coef(mod)[2]
b2
```

```
[1] 1.129626
```

The estimated variance of $\hat{\beta}_2$ is

```
v = vcov(mod)[2, 2]
v
```

---

[3]Let c be a constant and λ be a random variable, then $\text{Var}[c\lambda] = c^2\text{Var}[\lambda]$.
[4]Section 3.1.1

```
[1] 0.009859151
```

The natural logarithm of our estimate, $\log(\hat{\beta}_2)$, is

```
log(b2)
```

```
[1] 0.1218865
```

Using Equation 14.4, we compute the delta method estimate of $\text{Var}\left[\log(\hat{\beta}_2)\right]$, and we take its square root to obtain the standard error.

```
sqrt(v / b2^2)
```

```
[1] 0.08789924
```

To verify that this result is correct, we can use the `hypotheses()` function from the `marginaleffects` package. As noted in Chapter 4, `hypotheses()` can compute arbitrary functions of model parameters, along with standard errors.

```
library(marginaleffects)
hypotheses(mod, "log(b2) = 0")
```

Hypothesis	Estimate	Std. Error	z	Pr(>|z|)	2.5 %	97.5 %
log(b2)=0	0.122	0.0879	1.39	0.166	-0.0504	0.294

Reassuringly, the estimate and standard error that we computed "manually" for $\log(\hat{\beta}_2)$ are the same as the ones obtained "automatically" by calling `hypotheses()`.

14.1.2   Multivariate delta method

The univariate version of the delta method can be generalized to the multivariate case. Let $\mathcal{B} = \{\beta_1, \beta_2, \ldots, \beta_k\}$ be a vector of input parameters, and $\Theta = \{\theta_1, \theta_2, \ldots, \theta_n\}$ represent the vector-valued output of an h transformation. For example, $\mathcal{B}$ could be a vector of regression coefficients. Θ could be a single summary statistic, or it could be a vector of length n with one fitted value per observation in the dataset.[5]

Our goal is to compute standard errors for each element of Θ. To do this, we express the multivariate delta method in matrix notation

$$\text{Var}\left(h(\mathcal{B})\right) = J^T \cdot \text{Var}\left(\mathcal{B}\right) \cdot J, \tag{14.5}$$

---

[5]Chapter 5

where $h(\mathcal{B})$ is a function of the parameter vector $\mathcal{B}$ and $\mathrm{Var}(h(\mathcal{B}))$ is the variance of that function.

J is the Jacobian of h. The number of rows in that matrix is equal to the number of parameters in the input vector $\mathcal{B}$. The number of columns in J is equal to the number of elements in the output vector Θ, that is, equal to the number of transformed quantities of interest that h produces. The element in the ith row and j^{th} column of J is a derivative which characterizes the effect of a small change in β_i on the value of θ_j.

$$
J = \begin{bmatrix}
\frac{\partial \theta_1}{\partial \beta_1} & \frac{\partial \theta_2}{\partial \beta_1} & \cdots & \frac{\partial \theta_n}{\partial \beta_1} \\
\frac{\partial \theta_1}{\partial \beta_2} & \frac{\partial \theta_2}{\partial \beta_2} & \cdots & \frac{\partial \theta_n}{\partial \beta_2} \\
\vdots & \vdots & \ddots & \vdots \\
\frac{\partial \theta_1}{\partial \beta_k} & \frac{\partial \theta_2}{\partial \beta_k} & \cdots & \frac{\partial \theta_n}{\partial \beta_k}
\end{bmatrix}
$$

Now, let's go back to the linear model defined by Equation 14.1, with coefficients β_1, β_2, and β_3. Imagine that we want to estimate the difference between the 2nd and 3rd coefficients of this model. This difference can be expressed as the single output θ from a function h of three coefficients

$$
\theta = h(\beta_1, \beta_2, \beta_3) = \beta_2 - \beta_3, \tag{14.6}
$$

or

```
b = coef(mod)
b[2] - b[3]
```

```
[1] 0.8447632
```

To quantify the uncertainty around θ, we use the multivariate delta method formula. In Equation 14.5, $\mathrm{Var}(\mathcal{B})$ refers to the variance-covariance matrix of the regression coefficients, which we can extract using the vcov() function.

```
V = vcov(mod)
V
```

```
             (Intercept)            X            Z
(Intercept) 0.0094869866 0.0009058965 0.0002349717
X           0.0009058965 0.0098591507 0.0005353338
Z           0.0002349717 0.0005353338 0.0121275336
```

The J matrix is defined by reference to the h function. In Equation 14.6, h accepts three parameters as inputs and returns a single value. Therefore, the corresponding J matrix must have 3 rows and 1 column. We populate this matrix by taking partial derivatives of Equation 14.6 with respect to each of the input parameters.

$$J = \begin{bmatrix} \frac{\partial \theta}{\beta_1} \\ \frac{\partial \theta}{\beta_2} \\ \frac{\partial \theta}{\beta_3} \end{bmatrix} = \begin{bmatrix} 0 \\ 1 \\ -1 \end{bmatrix}$$

Having extracted Var($\mathcal{B}$) and defined an appropriate J for the quantity of interest θ, we now use Equation 14.5 to compute the standard error.

```
J = matrix(c(0, 1, -1), ncol = 1)
variance = t(J) %*% V %*% J
sqrt(variance)
```

[1] 0.1446237

We have thus estimated that the difference between β_2 and β_3 is 0.845, and that the standard error associated with this difference is 0.145. Again, checking this result against the output of the hypotheses() function shows that they are identical.

```
hypotheses(mod, hypothesis = "b2 - b3 = 0")
```

Hypothesis	Estimate	Std. Error	z	Pr(>|z|)	2.5 %	97.5 %
b2-b3=0	0.845	0.145	5.84	<0.001	0.561	1.13

14.1.3   Robust or clustered standard errors

The calculations in the previous sections relied on "classical" estimates of the variance-covariance matrix. These estimates are useful, but they rely on strong assumptions that rule out ubiquitous phenomena like autocorrelation, heteroskedasticity, and clustering.[6] Where such patterns occur, classical standard errors may inaccurately characterize the uncertainty in our estimates.

To see how the assumptions that underpin classical standard errors can be relaxed, let's consider a simple linear regression

$$Y = X\beta + \varepsilon,$$

---

[6]Autocorrelation is often an issue in time series estimation when the prediction errors we make for a unit of observation at time t are correlated to the prediction errors we make for the same unit at time $t + 1$. Clustering sometimes occurs when errors are similar for study participants drawn from well-defined groups (e.g., classrooms, neighborhoods, or provinces). Heteroskedasticity arises when the variance of the errors is inconsistent across observations, such as when our model makes bigger prediction errors for units with certain characteristics.

where Y is an $n \times 1$ vector of outcome; X is an $n \times k$ matrix of covariates, including an intercept; β is a $k \times 1$ vector of coefficients; ε is an $n \times 1$ vector of errors. We can estimate regression coefficients by ordinary least squares using the standard formula.

$$\hat{\beta} = (X^T X)^{-1}(X^T Y) \tag{14.7}$$

The usual way to compute the variance of $\hat{\beta}$ in this context is to use the "sandwich" formula, so-called because it includes a "meat" expression, squeezed between identical "bread" components. The classical variance estimate for ordinary least squares can thus be written as

$$Var(\hat{\beta}) = (X^T X)^{-1}(\sigma_\varepsilon^2 X^T X)(X^T X)^{-1}, \tag{14.8}$$

where σ_ε^2 can be estimated by taking the variance of estimated residuals.

To illustrate the use of these formulas, let's load the Thornton (2008) data and fit a linear regression model with one predictor and an intercept.

```
library(sandwich)
library(marginaleffects)
dat = get_dataset("thornton")
dat = na.omit(dat[, c("incentive", "outcome")])
mod = lm(outcome ~ incentive, data = dat)
summary(mod)
```

```
             Estimate Std. Error t value Pr(>|t|)
(Intercept)   0.33977    0.01695   20.04   <0.001
incentive     0.45106    0.01919   23.50   <0.001
```

The same estimates can be obtained manually by constructing a design matrix X, a response matrix Y, and by applying Equation 14.7.

```
X = cbind(1, dat$incentive)
Y = dat$outcome
solve(crossprod(X, X)) %*% crossprod(X, Y)
```

```
          [,1]
[1,] 0.3397746
[2,] 0.4510603
```

The variance-covariance matrix for these coefficients is obtained via Equation 14.8.

```
e = Y - predict(mod)
var_e = sum((e - mean(e))^2) / df.residual(mod)
bread = solve(crossprod(X))
meat = var_e * crossprod(X)
V = bread %*% meat %*% bread
V
```

```
             [,1]            [,2]
[1,]   0.0002874265 -0.0002874265
[2,]  -0.0002874265  0.0003684119
```

This is the same as the matrix returned by the vcov() function from base R.

```
vcov(mod)
```

```
                (Intercept)      incentive
(Intercept)   0.0002874265 -0.0002874265
incentive    -0.0002874265  0.0003684119
```

And the square root of the diagonal elements of this matrix are the standard errors printed in the model summary at the top of this section.

```
sqrt(diag(V))
```

```
[1] 0.01695366 0.01919406
```

One key assumption about classical standard errors is encoded explicitly in Equation 14.8. In that formula, and in the code that implements it, σ_ε^2 is a single number, treated as a constant that reflects the uncertainty for each observation uniformly. But in many applied contexts, the variance in errors is not uniform, that is, there is heteroskedasticity.[7]

One popular strategy to overcome with this challenge is to report "robust" standard errors, also known as "heteroskedasticity-consistent" or "Huber-White" standard errors. There are many types of robust standard errors available, designed to address different data issues (Zeileis et al., 2020). The code below shows one of the simplest ways to account for heteroskedasticity. It implements the same sandwich formula as in the classical case, but the "meat" component includes a matrix with squared residuals on the diagonal. This gives us more flexibility to account for changes in the variance of prediction errors across the sample.

```
meat = t(X) %*% diag(e^2) %*% X
bread %*% meat %*% bread
```

```
             [,1]            [,2]
[1,]   0.0003612364 -0.0003612364
[2,]  -0.0003612364  0.0004362886
```

---

[7] Footnote 6

A more convenient way to compute the same variance-covariance matrix is to use the vcovHC() function from the powerful sandwich package, which supports many models and estimators beyond linear regression.

```
vcovHC(mod, type = "HC0")
```

```
                 (Intercept)      incentive
(Intercept)   0.0003612364  -0.0003612364
incentive    -0.0003612364   0.0004362886
```

In all marginaleffects functions, there is a vcov argument that allows us to report various types of robust standard errors. When that argument is omitted, marginaleffects returns estimates with classical standard errors. To report heteroskedasticity consistent standard errors instead, we simply add vcov. For example, the next line of code returns an estimate of the average counterfactual comparison (risk difference) with heteroskedasticity-consistent standard error.

```
avg_comparisons(mod, vcov = "HC0")
```

Estimate	Std. Error	z	Pr($>$|z|)	2.5 %	97.5 %
0.451	0.0209	21.6	$<$0.001	0.41	0.492

The vcov argument accepts several string shortcuts like the one above. It can also take a variance-covariance matrix, a function that returns such a matrix, or a formula to specify clusters. The code block below shows a few examples, but readers are encouraged to consult the manual pages by typing ?hypotheses in R or help(hypotheses) in Python.

```
V = sandwich::vcovHC(mod)
avg_comparisons(mod, vcov = V)          # heteroskedasticity-consistent
avg_comparisons(mod, vcov = ~ village)  # clustered by village
```

In sum, we have seen that the delta method is a simple, fast, and flexible approach to obtain standard errors for a vast array of quantities of interest. It is the default approach implemented by all the functions in the marginaleffects package. By combining the delta method with robust estimates of the variance-covariance matrix, analysts can report uncertainty estimates that account for common data features like heteroskedasticity or clustering.

14.2 Bootstrap

The bootstrap is a method pioneered by Bradley Efron and colleagues in the 1970s and 1980s (Efron and Tibshirani, 1994). It is a powerful strategy to quantify uncertainty about quantities of interest.[8]

Consider T_n, a statistic we wish to compute. T_n is a function of an outcome Y, some regressors $\mathbf{X}$, and the unknown population distribution function F:

$$T_n = T_n((Y_1, \mathbf{X}_1), \ldots, (Y_n, \mathbf{X}_n), F)$$

The statistic of interest T_n follows a sampling distribution G, which depends on the unknown population distribution F. If we knew F, then G would immediately follow. The key idea of the bootstrap is to use simulations to approximate F, and then to use that approximation when quantifying uncertainty around the quantity of interest.

How do we approximate F in practice? Since the 1980s, many variations on bootstrap simulations have been proposed. A survey of this rich literature lies outside the scope of this book, but it is instructive to consider one of the simplest approaches: the non-parametric bootstrap algorithm. The idea is to build several simulated datasets by drawing repeatedly from the observed sample. With those datasets in hand, we can estimate the statistic of interest multiple times, and characterize its sampling distribution.

To illustrate, let's consider a simple linear model fit to the Thornton (2008) data. The response is `outcome`, which is a binary variable equal to 1 for study participants who sought to learn their HIV status, and 0 for the others. There are three coefficients, each representing an age category.

```
dat = get_dataset("thornton")
mod = lm(outcome ~ agecat - 1, data = dat)
coef(mod)
```

```
    agecat<18 agecat18 to 35     agecat>35
    0.6718750      0.6787004     0.7277354
```

Our substantive goal is to determine if the proportion of individuals with `outcome=1` is higher in the 35+ age group than in the 18-to-35 bracket. In other words, the quantity of interest is the difference between the 2nd and 3rd coefficients of the model. For convenience, we define a `statistic()` function to compute this difference.

---

[8]The notation and ideas in this section follow Hansen (2022b).

```
statistic = \(model) coef(model)[3] - coef(model)[2]
statistic(mod)
```

```
[1] 0.04903501
```

Now, let's compute a bootstrap confidence interval around this quantity. This can be achieved in five steps:

1. Sample rows with replacement from the original data.
2. Fit the model to the new dataset.
3. Compute the statistic using the new fitted model's parameter estimates.
4. Repeat steps 1 through 3 many times.
5. Use quantiles of the bootstrap distribution to build a confidence interval around the statistic of interest.

The following code implements these five steps.

```
sample_fit_compute = function() {
  # Step 1: Sample
  index = sample(1:nrow(dat), size = nrow(dat), replace = TRUE)
  resample = dat[index, ]

  # Step 2: Fit
  mod_resample = lm(outcome ~ agecat - 1, data = resample)

  # Step 3: Compute
  statistic(mod_resample)
}

# Step 4: Repeat
set.seed(48103)
boot_dist = replicate(1000, sample_fit_compute())

# Step 5: Quantiles
ci = quantile(boot_dist, prob = c(0.025, 0.975))
ci
```

```
      2.5%       97.5%
0.01435133 0.08701632
```

This analysis suggests that the estimated difference between the 3rd and the 2nd coefficient is 0.049, with a 95% confidence interval of [0.014, 0.087].

The `inferences()` function from `marginaleffects` makes this kind of analysis much easier, by providing a consistent interface to the `boot`, `rsample`, and `fwb` packages (Canty and Ripley, 2022; Frick et al., 2022; Greifer, 2022). This empowers users to apply a wide variety of bootstrap resampling schemes. To use this functionality, we call any of the core `marginaleffects` functions, and pass its output to `inferences()`.[9]

```
hypotheses(mod, "b3 - b2 = 0") |>
  inferences(method = "boot", R = 1000)
```

Hypothesis	Estimate	2.5 %	97.5 %
b3-b2=0	0.049	0.0156	0.0846

The results are close those obtained manually, but they vary a little bit due to the random nature of the procedure and minor differences in the resampling scheme.

Note that the same workflow works for other `marginaleffects` functions, such as `avg_comparisons()`.

```
avg_comparisons(mod) |> inferences(method = "boot")
```

Contrast	Estimate	2.5 %	97.5 %
18 to 35 − <18	0.00683	−0.05439	0.0668
>35 − <18	0.05586	−0.00393	0.1151

Bootstrapping is a versatile approach to estimate the sampling distribution of a statistic. It is widely applicable, even if we do not know the underlying distribution of the quantities of interest, making it a powerful tool for deriving standard errors and confidence intervals. However, bootstrapping can be computationally intensive, particularly for large datasets or complex models. In addition, selecting an adequate resampling scheme can be challenging when the data-generating process is complex.

14.3   Simulation

Yet another strategy to estimate the uncertainty around quantities of interest is simulation-based inference. This method is an alternative to traditional analytical approaches, leveraging the power of computational simulations to estimate the variability of a given function of parameters. One of the earliest

---

[9]Additional arguments can be passed to control functions from the bootstrap packages, by directly inserting them in the `inferences()` call. See `?boot::boot`, `?rsample::rsample`, and `?fwb::fwb`

applications of this approach can be found in Krinsky and Robb (1986), who use simulations to quantify the sampling uncertainty around estimates of elasticities in a regression modeling context. King et al. (2000) subsequently popularized the strategy and its implementation in the `clarify` software.[10]

The foundational assumption of this approach is that a regression-based estimator is normal. This allows us to leverage the properties of the multivariate normal distribution to simulate many sets of coefficients, and many quantities of interest. This can be done in 3 steps:

1. Draw many sets of parameters from a multivariate normal distribution, with mean equal to the estimated vector of coefficients, and variance equal to the estimated covariance matrix.
2. Compute the statistic using each set of simulated coefficients.
3. Build a confidence interval for the quantity of interest by taking quantiles of the simulated quantities.

As Rainey (2024) notes, this process can be thought of as an informal Bayesian posterior simulation. While it does not strictly use Bayesian methods, it shares a similar outlook, using the distribution of simulated parameters to infer the uncertainty around transformed quantities. By repeatedly sampling from the parameter space, we can gain a clearer picture of the possible variability.

The code below implements a simple version of this strategy.

```
# Step 1: Simulate coefficients
library(mvtnorm)
set.seed(48103)
coef_sim = rmvnorm(1000, coef(mod), vcov(mod))

# Step 2: Compute the quantity of interest
statistics = coef_sim[, 3] - coef_sim[, 2]

# Step 3: Build confidence intervals as quantiles
ci = quantile(statistics, probs = c(0.025, 0.975))
ci
```

```
      2.5%       97.5%
0.01343765 0.08461654
```

Note that the simulation-based confidence interval is slightly different from those obtained via bootstrap or delta method, but it has the same general magnitude.

We can obtain similar results with the **inferences()** function.

---

[10] `clarify` is now available for R (Greifer et al., 2024).

```
hypotheses(mod, "b3 - b2 = 0") |>
  inferences(method = "simulation", R = 1000)
```

Hypothesis	Estimate	2.5 %	97.5 %
b3-b2=0	0.049	0.015	0.0871

```
avg_comparisons(mod, variables = "agecat") |>
  inferences(method = "simulation", R = 1000)
```

Contrast	Estimate	2.5 %	97.5 %
18 to 35 − <18	0.00683	−0.05112	0.0599
>35 − <18	0.05586	0.00105	0.1134

Simulation-based inference is easy to implement, convenient, and it applies to a broad range of models. It trades the analytic approximation of the delta method for a numerical approximation through simulation. Although simulation-based inference can be more computationally expensive than the delta method, it is usually faster than bootstrapping, which requires refitting the model multiple times.

14.4   Conformal prediction

In Chapter 5, we used the `predictions()` function to compute confidence intervals around fitted values. These intervals are designed to characterize the uncertainty about the model-based expected value of the response.

One common misunderstanding is that these confidence intervals should be calibrated to cover a certain percentage of unseen data points. This is not the case. In fact, a confidence interval built using the delta method will typically cover a much smaller share of out-of-sample observations than its nominal size would suggest.

To see this, let's conduct a simulation with two samples of 25 observations drawn randomly from a normal distribution with mean π. The Y_{train} set is used to fit a model, and the set Y_{test} is used to check the accuracy of the model's predictions. The model we fit is a linear regression with only an intercept, and we use the `predictions()` function to compute a 90% confidence interval around the expected value of the response. Finally, we check how many observations in the test set are covered by this interval.

```
simulation = function(...) {
  # Draw random training and test sets
  Y_train = rnorm(25, mean = 3.141593)
  Y_test = rnorm(25, mean = 3.141593)

  # Fit an intercept-only linear model
  m = lm(Y_train ~ 1)

  # Predictions with 90% confidence interval
  p = predictions(m, conf_level = .90)

  # Does the confidence interval cover the true mean?
  coverage_mean = mean(p$conf.low < 3.141593 & p$conf.high > 3.141593)

  # Does the confidence interval cover out of sample observations?
  coverage_test = mean(p$conf.low < Y_test & p$conf.high > Y_test)

  out = c(
    "Mean coverage" = coverage_mean,
    "Out-of-sample coverage" = coverage_test)
  return(out)
}
```

To draw an overall portrait of performance, we repeat this experiment 1000 times, and report the average coverage across experiments.

```
set.seed(48103)
coverage = rowMeans(replicate(1000, simulation()))
coverage
```

```
       Mean coverage Out-of-sample coverage
          0.88100                0.24668
```

The confidence intervals produced by **predictions()** cover the true mean of the outcome nearly 90% of the time, which matches the significance level that we requested using the **conf_level** argument. However, the confidence intervals only cover out-of-sample observations 25% of the time. Clearly, if we want an interval to cover a pre-specified share of unseen data points, we cannot rely on *confidence* intervals. We must compute *prediction* intervals instead.

How can we obtain prediction intervals with adequate coverage? In their excellent tutorial, Angelopoulos and Bates (2022) introduce "conformal prediction" as

"a user-friendly paradigm for creating statistically rigorous uncertainty sets/intervals for the predictions of such models. Critically, the sets are valid in a distribution-free sense: they possess explicit, non-asymptotic guarantees even without distributional assumptions or model assumptions."

These are extraordinary claims which deserve to be repeated and empha-
sized. Conformal prediction can offer well-calibrated intervals, that cover a
known share of out-of-sample data points, in finite samples, without making
distributional assumptions, and even if the prediction model is misspecified.

There are three main caveats to this. First, the conformal prediction algorithms
implemented in `marginaleffects` are designed for exchangeable data.[11] They
do not offer coverage guarantees in contexts where exchangeability is violated,
such as in time series data, when there is spatial dependence between observations, or when there is distribution drift between the training and test data.
Second, these algorithms offer marginal coverage guarantees, that is, they
guarantee that a random test point will fall within the interval with a given
probability. Such intervals may not be well calibrated locally, in different strata
of the predictors.[12] Third, the width of the conformal prediction interval will
typically depend on the quality of the prediction model and of the function that
we use to characterize the "quality" of predictions (i.e., the "score" function).[13]

A detailed technical treatment of conformal prediction lies outside the scope
of this book, but we can form some intuition by considering two prediction
tasks: numeric and categorical outcomes.

14.4.1   Numeric outcome

This introduction to conformal prediction adopts a code-first rather than a
theory-heavy approach. To make things concrete, we will consider a dataset
which contains demographic information on over 1 million members of the US
military, including variables such as grade, branch, gender, race, and rank. Our
prediction task involves using a linear model to predict the rank of military
personnel based on these individuals' characteristics.

```
library(marginaleffects)
dat = get_dataset("military")
head(dat)
```

```
# A data frame: 6 x 8
  rownames grade    branch gender race     hisp  rank officer
*    <int> <chr>    <chr>  <chr>  <chr>    <lgl> <int>   <dbl>
1        1 officer  army   male   ami/aln  TRUE     2       1
2        2 officer  army   male   ami/aln  TRUE     2       1
```

---

[11] The usual "independent and identically distributed" assumption is a more restrictive
special case of exchangeability.

[12] Different algorithms have recently been proposed to offer class-conditional coverage guarantees (Ding et al., 2023). These algorithms are not currently implemented in
`marginaleffects`, but they are on the software development roadmap.

[13] Note that the score functions implemented in `marginaleffects` simply take the residual—
or difference between the observed outcome and predicted value. This means that the `type`
argument must ensure that observations and predictions are on commensurable scales
(usually `type="response"` or `type="prob"`).

3	3 officer army	male	ami/aln TRUE	5	1
4	4 officer army	male	ami/aln TRUE	5	1
5	5 officer army	male	ami/aln TRUE	5	1
6	6 officer army	male	ami/aln TRUE	5	1

There are many algorithms available to build prediction intervals. For illustration, we will use one of the simplest: split conformal prediction. The first step of this strategy is to split the dataset randomly into three parts.[14] The training set is used to fit the prediction model. The calibration set is used by the conformal prediction algorithm to determine the appropriate width of prediction intervals. The test set is used to evaluate the accuracy of out-of-sample predictions.

```
set.seed(48103)
idx = sample(
  c("train", "calibration", "test"),
  size = nrow(dat),
  replace = TRUE)
dat = split(dat, idx)
```

Now we use the training data to fit a linear regression model. Our goal is to predict the rank of an individual member of the military, based on four predictors.

```
mod = lm(rank ~ grade + branch + gender + race, data = dat$train)
summary(mod)
```

	Estimate	Std. Error	t value	Pr(>|t|)
(Intercept)	5.900195	0.020494	287.893	<0.001
gradeofficer	-1.288080	0.006967	-184.880	<0.001
gradewarrant officer	-2.167464	0.021205	-102.213	<0.001
brancharmy	-0.192071	0.006500	-29.551	<0.001
branchmarine corps	-0.705205	0.008311	-84.848	<0.001
branchnavy	-0.064200	0.007314	-8.777	<0.001
gendermale	0.286665	0.007123	40.243	<0.001
raceasian	0.430081	0.022936	18.751	<0.001
raceblack	0.652504	0.019943	32.718	<0.001
racemulti	-0.584264	0.026247	-22.260	<0.001
racep/i	-0.025871	0.036875	-0.702	0.483
raceunk	1.261640	0.022067	57.173	<0.001
racewhite	0.401818	0.019304	20.816	<0.001

Split conformal inference leverages the assumption that observations in the calibration set are *exchangeable* for those in the test set. If, in that sense, the calibration set is similar to the test set, then it seems reasonable to expect that the prediction errors we make in one will be similar to the prediction errors

---

[14]In this case, the three datasets have approximately the same number of rows, but this need not be the case.

we make in the other.[15] With that in mind, we can study the calibration set to get a sense of how wide the prediction intervals should be.

The code below does three things: (1) use the fitted linear model to make predictions in the calibration set, (2) compute absolute residuals for each observation in the calibration set, and (3) plot the distribution of those residuals. The vertical dashed line shows the 95th quantile of absolute residuals.

```
# Predictions in the calibration set
yhat_calib = predict(mod, newdata = dat$calib)

# Absolute residuals in the calibration set
resid_calib = abs(dat$calib$rank - yhat_calib)

# Plot
hist(resid_calib, breaks = 50, main = NULL, xlab = "|e|")
abline(v = quantile(resid_calib, 0.95), lty = 2)
```

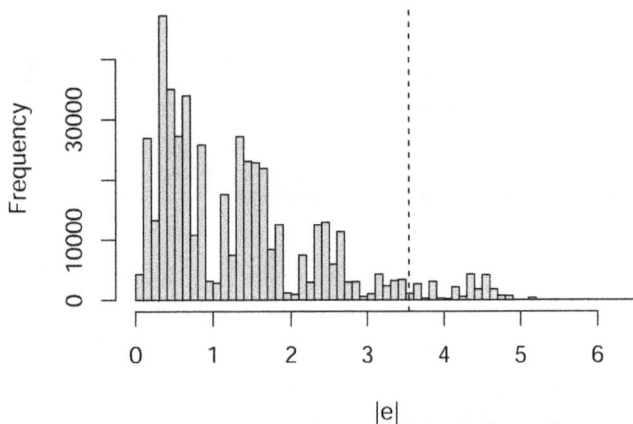

Figure 14.1: Distribution of absolute residuals in the calibration set, with the 95th quantile identified by a dashed vertical line.

If the calibration and test sets are exchangeable, then it feels reasonable to expect that 95% of the absolute prediction errors that our model will make in the test set will fall to the left of the dashed line in Figure 14.1. Thus, if we want to construct a prediction interval that covers at least 95% of the points in the test set, we can define a constant d equal to the smallest absolute residual above the 95th percentile, and build prediction intervals as $\hat{Y}_i \pm d$.

---

[15]Prediction errors in the training set are usually smaller than errors in either the calibration or test sets, because of overfitting.

```
# Predictions in the test set
yhat_test = predict(mod, newdata = dat$test)

# Measure the width of prediction intervals
q95 = quantile(resid_calib, probs = 0.95)
d = min(resid_calib[resid_calib > q95])

# Build prediction intervals
conformal_lo = yhat_test - d
conformal_hi = yhat_test + d
```

Finally, we can check how many of the unseen data points in the test set are covered by the split conformal prediction sets.

```
covered = dat$test$rank > conformal_lo & dat$test$rank < conformal_hi
mean(covered)
```

```
[1] 0.9515136
```

We find that 95.2% of observations in the test set are covered. This is very close to our target coverage rate of 95%.

Instead of doing all that by hand, we can use the `inferences()` function from the `marginaleffects` package, which yields identical results.

```
p = predictions(mod) |>
  inferences(
    method = "conformal_split",
    conformal_calibration = dat$calib,
    conformal_test = dat$test)
```

```
covered = p$rank > p$pred.low & p$rank < p$pred.high
mean(covered)
```

```
[1] 0.9515136
```

14.4.2 Categorical outcome

Now, let's illustrate an analogous strategy in the context of a classification task. We use a multinomial logit model to predict what branch of the military each observation belongs to: air force, army, marine corps, or navy.

```
library(nnet)
mod = multinom(branch ~ gender * race, data = dat$train, trace = FALSE)
```

The key difference with the previous section is that the outcome is now nominal, rather than numeric. For each observation in the test set, we want to predict a set of classes that should include the true value with a specified probability. For some individual, the model could thus predict a set of classes {navy, air force} that, we hope, includes the true branch to which the individual belongs.

Recall that conformal prediction requires us to quantify the quality of predictions in a calibration dataset. Instead of using a residuals-based approach as in the previous section, we measure the quality of class predictions using a "softmax score," derived from the probability that the model assigns to the true (observed) class. Also, rather than preparing a held-out calibration set as in Section 14.4.1, we use an alternative conformal algorithm based on cross-validation.[16] These changes can be made easily by modifying the arguments of the `inferences()` function.

```
set.seed(48103)
p = predictions(mod, conf_level = .8) |>
    inferences(
        R = 5,
        method = "conformal_cv+",
        conformal_score = "softmax",
        conformal_test = dat$test)
```

The `inferences()` returns a data frame with one row per observation in the test set. The `pred.set` column from that data frame includes one list of possible classes, for each observation.

Consider the last individual in our test set. If we call `predictions()` on the data for that individual, we find that the most likely **branch** value "army."

Group	Estimate	Std. Error	z	Pr(>|z|)	2.5 %	97.5 %
air force	0.365	0.00746	48.9	<0.001	0.3503	0.380
army	0.422	0.00765	55.2	<0.001	0.4073	0.437
marine corps	0.100	0.00466	21.6	<0.001	0.0912	0.109
navy	0.112	0.00489	23.0	<0.001	0.1028	0.122

```
predictions(mod, newdata = tail(dat$test, 1))
```

However, making that single point prediction would be misleading, because that individual actually belongs to a different branch.

```
tail(dat$test$branch, 1)
```

```
[1] "air force"
```

---

[16]The split conformal algorithm would also work in this context. We use cross-validation simply for illustration.

Thankfully, our conformal prediction set included more than just a point prediction. Indeed, we see that the prediction set associated to that individual includes the most likely (but incorrect) branch, but also the correct result.

```
tail(p, 1)
```

Pred Set 90.0 %
air force, army

More generally, we can check if the individual prediction sets include the true values of **branch** in the test set, using the code below.

```
mean(unlist(Map(\(o, p) o %in% p, p$branch, p$pred.set)))
```

```
[1] 0.8202748
```

As expected, conformal prediction sets include the correct value of about 80% of the time.

In conclusion, this chapter has explored various methods to quantify uncertainty in statistical analysis, including the delta method, bootstrap, simulation-based inference, and conformal prediction. Each method offers unique advantages and is suitable for different scenarios, depending on the assumptions one is willing to make and on the computational resources available. The delta method provides fast results using linear approximations, while the bootstrap and simulation-based inference leverage computational power to draw from re-sampled or simulated results. Conformal prediction offers distribution-free prediction intervals which, under certain conditions, can be expected to cover a specified share of unseen data points. Understanding these methods equips analysts with a comprehensive toolkit to address uncertainty in their analyses.

Part IV

Back matter

Appendix I: Online content

The `marginaleffects` package is a powerful and versatile tool for statistical analysis, with features that extend well beyond what we can cover in this book. These features are documented in detail on the package website, along with many more case studies and examples: https://marginaleffects.com. What follows is a partial list of the ever growing resources available on the website. At that address, you will also find a discussion forum for questions and support, the complete source code for both R and Python implementations, useful links to related resources, fun statistical memes, and much more.

- Help & FAQ
- Bayesian inference
- Conjoint experiments
- Contrasts
- Elasticity
- Extending `marginaleffects`
- Generalized additive models
- Interrupted time series
- Interval tests
- Inverse probability weighting
- Joint hypothesis tests
- Marginal means
- Matching
- Missing data
- Multilevel regression
- Survival analysis
- Targeted maximum likelihood estimation, etc.
- etc.

DOI: 10.1201/9781003560333-I

Appendix II: Python

This chapter collects **Python** code equivalents to nearly all of the R commands shown in this book. The same code is reproduced, with full output, on the marginaleffects.com website. The small number of features that are not yet available in the **Python** version of **marginaleffects** are listed on the roadmap at the end of this appendix.

Note that, by default, **marginaleffects** commands return data frames in the Polars format. These objects are easy to convert to Pandas, Numpy, or other formats using an appropriate `.to_*()` method. For example, calling the `get_dataset()` function returns a Polars data frame that we can convert to Pandas with `.to_pandas()`. Recall that Python uses 0-based indexing whereas R uses 1-based indexing.

```
import polars as pl
import numpy as np
from marginaleffects import *
from plotnine import *
from scipy.stats import norm
from statsmodels.formula.api import logit, ols

dat = get_dataset("thornton")
dat = dat.to_pandas()
dat.head()
```

1. Who is this book for?

Section 1.4

```
dat = get_dataset("Titanic", "Stat2Data")
dat[:6, ["Name", "Survived", "Age"]]
get_dataset(search = "Titanic")
get_dataset("Titanic", "Stat2Data", docs = True)
```

3. Conceptual framework

Section 3.2

```
rng = np.random.default_rng(seed=48103)
N = 10
dat = pl.DataFrame({
  "Num": rng.normal(loc = 0, scale = 1, size = N),
  "Bin": rng.binomial(n = 1, p = 0.5, size = N),
  "Cat": rng.choice(["A","B","C"], size = N)
})
```

Empirical grid

```
dat
```

Interesting grid

```
datagrid(Bin=[0,1], newdata = dat)
datagrid(
  Num=[dat["Num"].max(), dat["Num"].min()],
  Bin=dat["Bin"].mean(),
  Cat=dat["Cat"].unique(),
  newdata = dat)
```

Representative grid

```
datagrid(grid_type = "mean_or_mode", newdata = dat)
```

Balanced grid

```
datagrid(grid_type = "balanced", newdata = dat)
```

Counterfactual grid

```
g = datagrid(
  Bin=[0,1],
  grid_type = "counterfactual",
  newdata = dat
)
g.shape
g.filter(pl.col("rowidcf")<3)
```

4. Hypothesis and equivalence tests

```
dat = get_dataset("thornton")
dat.head(6)
mod = ols("outcome ~ agecat - 1",
  data=dat.to_pandas()).fit()
mod.params
dat.group_by("agecat").agg(pl.col("outcome").mean())
```

Section 4.1

Choice of null hypothesis

```
hypotheses(mod, hypothesis = 0.5)
b = mod.params.iloc[0]
se = np.sqrt(np.diag(mod.cov_params()))[0]
z = (b - 0.5) / se
(norm.cdf(-np.abs(z))) * 2
```

Linear and non-linear hypothesis tests

```
hypotheses(mod, hypothesis = "b2 - b0 = 0")
hypotheses(mod, hypothesis = "b2 / b0 = 1")

import numpy as np
hypotheses(mod, hypothesis = "b1**2 * np.exp(b0) = 0")
hypotheses(mod, hypothesis = "b0 - (b1 * b2) = 2")

hypotheses(mod, hypothesis = "difference ~ reference")
hypotheses(mod, hypothesis = "ratio ~ sequential")
```

Section 4.2

```
hypotheses(mod, hypothesis = "b2 - b1 = 0")
hypotheses(mod, hypothesis = "b2 - b1 = 0",
  equivalence = [-0.05, 0.05])["p_value_equiv"]
```

5. Predictions

Section 5.1

```
dat = get_dataset("thornton")
dat = dat.drop_nulls(subset = ["incentive"])
mod = logit("outcome ~ incentive + agecat",
  data=dat.to_pandas()).fit()
b = mod.params

linpred_treatment_younger = b.iloc[0] + b.iloc[3] * 1 + \
  b.iloc[1] * 1 + b.iloc[2] * 0
linpred_control_older = b.iloc[0] + b.iloc[3] * 0 + \
  b.iloc[1] * 0 + b.iloc[2] * 1

logistic = lambda x: 1 / (1 + np.exp(-x))
logistic(linpred_treatment_younger)
logistic(linpred_control_older)

grid = pl.DataFrame({
    "agecat": ["18 to 35", ">35"],
    "incentive": [1, 0]
})

predictions(mod, newdata = grid)
```

Section 5.2

```
mod = logit("outcome ~ incentive + agecat + distance",
  data=dat.to_pandas()).fit()
```

Empirical grid

```
p = predictions(mod)
p.shape
p.columns
p[:4, "estimate"]
```

Interesting grid

```
datagrid(agecat = "18 to 35", incentive = [0,1], model = mod)
predictions(mod,
  newdata=datagrid(agecat = "18 to 35", incentive = [0,1]))
predictions(mod,
  newdata=datagrid(
    distance = 2,
    agecat = dat["agecat"].unique(),
    incentive = dat["incentive"].max())))
```

Representative grid

```
predictions(mod, newdata="mean")
```

Balanced grid

```
predictions(mod,
  newdata = datagrid(
    agecat = dat["agecat"].unique(),
    incentive = dat["incentive"].unique(),
    distance = dat["distance"].mean())))
predictions(mod, newdata = "balanced")
```

Counterfactual grid

```
p = predictions(mod, variables = {"incentive": [0, 1]})
p.shape
p = pl.DataFrame({
    "Control": p.filter(pl.col("incentive") == 0)["estimate"],
    "Treatment": p.filter(pl.col("incentive") == 1)["estimate"]
})
plot = (
  ggplot(p, aes(x="Control", y="Treatment")) +
  geom_point() +
  geom_abline(
    intercept=0, slope=1, linetype="--", color="grey") +
  labs(
    x="Pr(outcome=1) when incentive = 0",
    y="Pr(outcome=1) when incentive = 1") +
  xlim(0, 1) +
  ylim(0, 1) +
  theme_minimal()
)
plot.show()
```

Section 5.3

```
p = predictions(mod)
p["estimate"].mean()
avg_predictions(mod)
avg_predictions(mod, by="agecat")
avg_predictions(mod, newdata="balanced", by="agecat")
dat["incentive"].value_counts()
avg_predictions(mod, by="incentive")
avg_predictions(mod,
  variables={"incentive": [0, 1]},
  by="incentive")
```

Section 5.5

Null hypothesis tests

```
p = avg_predictions(mod, by="agecat")
p["estimate"][2]-p["estimate"][1]
avg_predictions(mod, by="agecat", hypothesis = "b2 - b1 = 0")
avg_predictions(mod, by="agecat",
  hypothesis = "difference ~ sequential")
avg_predictions(mod, by="agecat",
  hypothesis = "difference ~ reference")
avg_predictions(mod, by=["incentive","agecat"])
avg_predictions(mod, by=["incentive","agecat"],
  hypothesis = "difference ~ sequential | incentive")
```

Equivalence tests

```
avg_predictions(mod,
  by="agecat",
  hypothesis = "b2 - b0 = 0",
  equivalence = [-0.1, 0.1])
```

Section 5.6

Unit predictions

```
p = predictions(mod)
p = p.with_columns(pl.col("incentive").cast(pl.Utf8))
plot = (
  ggplot(p, aes(x="estimate", fill="incentive")) +
  geom_histogram() +
  labs(x="Pr(outcome=1) when incentive = 0"))
plot = (
  ggplot(p, aes(x="estimate", colour="incentive")) +
  geom_line(stat='ecdf'))
```

Marginal predictions

```
avg_predictions(mod, by="incentive")
plot_predictions(mod, by="incentive").show()
plot_predictions(mod, by=["incentive", "agecat"]).show()
plot_predictions(mod, by="incentive",
  newdata = "balanced", draw = False)
```

Conditional predictions

```
plot_predictions(mod, condition="distance").show()
plot_predictions(mod, condition=["distance","incentive"]).show()
plot_predictions(mod,
  condition=["distance","incentive", "agecat"]
).show()
plot_predictions(mod,
  condition={"distance" : None, "agecat" : ">35", "incentive" : 0 }
).show()
```

Customization

```
plot_predictions(mod, by="incentive", draw = False)
```

6. Counterfactual comparisons

```
dat = get_dataset("thornton")
mod = logit("outcome ~ incentive * (agecat + distance)",
  data=dat.to_pandas()).fit()
mod.summary()
```

Section 6.1.1

```
grid = pl.DataFrame({
    "distance": 2,
    "agecat": ["18 to 35"],
    "incentive": 1
})

g_treatment = grid.with_columns(pl.lit(1).alias("incentive"))
g_control = grid.with_columns(pl.lit(0).alias("incentive"))
p_treatment = mod.predict(g_treatment.to_pandas())
p_control = mod.predict(g_control.to_pandas())
p_treatment - p_control

comparisons(mod, variables = "incentive", newdata = grid)
```

Section 6.1.2

```
comparisons(mod,
  variables = "incentive",
  comparison = "ratio",
  hypothesis = 1,
  newdata = grid)

comparisons(mod,
  variables = "incentive",
  comparison = "lift",
  hypothesis = 1,
  newdata = grid)
```

Section 6.2

Focal variables.

```python
comparisons(mod,
  variables="incentive",
  newdata=grid)
comparisons(mod,
  variables={"incentive" : [1, 0]},
  newdata=grid)
comparisons(mod,
  variables="agecat",
  newdata=grid)
comparisons(mod,
  variables={"agecat" : ["18 to 35", ">35"]},
  newdata=grid)
```

```python
# Increase of 5 units
comparisons(mod, variables={"distance": 5}, newdata=grid)
# Increase of 1 standard deviation
comparisons(mod, variables={"distance": "sd"}, newdata=grid)
# Change between specific values
comparisons(mod, variables={"distance" : [0, 3]}, newdata=grid)
# Change across the interquartile range
comparisons(mod, variables={"distance" : "iqr"}, newdata=grid)
# Change across the full range
comparisons(mod, variables={"distance": "minmax"}, newdata=grid)
```

Cross-comparisons

```python
comparisons(mod,
  variables = ["incentive", "distance"],
  cross = True,
  newdata = grid)
```

Adjustment variables

```python
comparisons(mod, variables = "incentive")
cmp = comparisons(mod)
cmp.shape
```

```
comparisons(mod,
  variables = "incentive", newdata = datagrid(
    agecat = dat["agecat"].unique(),
    distance = dat["distance"].mean())))
comparisons(mod, variables = "incentive", newdata = "mean")
comparisons(mod, variables = "incentive", newdata = "balanced")
```

Section 6.3

```
avg_comparisons(mod, variables = "incentive")
cmp = comparisons(mod, variables = "incentive")
cmp["estimate"].mean()
avg_comparisons(mod, variables = "incentive", by = "agecat")
avg_comparisons(mod,
  variables = "incentive",
  newdata = dat.filter(pl.col("incentive") == 1))

penguins = get_dataset("penguins", "palmerpenguins") \
  .drop_nulls(subset = "species")
penguins \
  .select("species", "body_mass_g", "flipper_length_mm") \
  .group_by("species").mean()
fit = ols(
  "flipper_length_mm ~ body_mass_g * species",
  data = penguins.to_pandas()).fit()
avg_predictions(fit, by = "species")
avg_predictions(fit, variables = "species", by = "species")
avg_comparisons(fit, variables = {"species": "sequential"})
```

Section 6.5

```
avg_comparisons(mod, variables = "incentive", by = "agecat")
avg_comparisons(mod, variables = "incentive", by = "agecat",
  hypothesis = "b0 - b2 = 0")
```

Section 6.6

```
avg_comparisons(mod,
  variables = "incentive",
  by = "agecat")
plot_comparisons(mod,
  variables = "incentive",
  by = "agecat").show()
plot_comparisons(mod,
  variables = "incentive",
  condition = "distance").show()
plot_comparisons(mod,
  variables = "incentive",
  condition = ["distance", "agecat"]).show()
```

7. Slopes

```
import polars as pl
import numpy as np
from marginaleffects import *
from plotnine import *
from statsmodels.formula.api import logit, ols

np.random.seed(48103)
N = int(1e6)
X = np.random.normal(loc=0, scale=2, size=N)
p = 1 / (1 + np.exp(-(-1 + 0.5 * X)))
Y = np.random.binomial(n=1, p=p)
dat = pl.DataFrame({"Y": Y, "X": X})
mod = logit("Y ~ X", data=dat.to_pandas()).fit()
mod.params
slopes(
  mod,
  variables="X",
  newdata=datagrid(X = [-5, 0, 10]))
```

Section 7.2

```
dat = get_dataset("thornton")
mod = logit(
  "outcome ~ incentive * distance * I(distance**2)",
  data = dat.to_pandas()).fit()
slopes(
  mod,
  variables = "distance",
  newdata = datagrid(incentive = 1, distance = 1))
slopes(
  mod,
  variables = "distance",
  newdata = "mean")
slopes(mod, variables = "distance")
```

Section 7.3

```
avg_slopes(mod, variables = "distance")
avg_slopes(mod, variables = "distance", by = "incentive")
```

Section 7.5

```
avg_slopes(mod, variables = "distance", by = "incentive")
avg_slopes(mod,
  variables = "distance",
  by = "incentive",
  hypothesis = "b0 - b1 = 0")
```

Section 7.6

```
plot_predictions(mod, condition = "distance").show()
plot_slopes(mod, variables = "distance",
  condition = "distance").show()
plot_slopes(mod,
  variables = "distance",
  condition = ["distance", "incentive"]).show()
plot_slopes(mod,
  variables = "distance",
  by = "incentive").show()
```

8. Causal inference with G-computation

Section 8.1

```
dat = get_dataset("lottery")
dat = dat.filter((pl.col('win_big') == 1) | (pl.col('win') == 0))
mod = smf.ols("earnings_post_avg ~ win_big * (
  tickets + man + work + age + education + college + year +
  earnings_pre_1 + earnings_pre_2 + earnings_pre_3)",
  data=dat).fit()
mod.summary()


d0 = dat.with_columns(pl.lit(0).alias('win_big'))
d1 = dat.with_columns(pl.lit(1).alias('win_big'))
p0 = predictions(mod, newdata = d0)
p1 = predictions(mod, newdata = d1)
dat[5, 'win_big']
p0[5, 'estimate']
p1[5, 'estimate']


p0.select('estimate').mean()
p1.select('estimate').mean()
avg_predictions(mod, variables = "win_big", by = "win_big")


avg_comparisons(mod, variables = "win_big")
avg_comparisons(mod, variables = "win_big",
  newdata=dat.filter(pl.col('win_big') == 1))
avg_comparisons(mod, variables = "win_big",
  newdata=dat.filter(pl.col('win_big') == 0))
```

Section 8.2

```
avg_comparisons(mod, variables="win_big", by="work")
```

9. Experiments

Section 9.1

```
dat = get_dataset("thornton")
mod = ols("outcome ~ incentive", data = dat.to_pandas()).fit()
mod.params
avg_comparisons(mod, variables = "incentive", vcov = "HC2")


mod = ols("outcome ~ incentive * (age + distance + hiv2004)",
  data=dat.to_pandas()).fit()
mod.params
avg_comparisons(mod, variables = "incentive", vcov = "HC2")
```

Section 9.2

```
dat = get_dataset("factorial_01")
mod = ols("Y ~ Ta + Tb + Ta:Tb", data = dat.to_pandas()).fit()
mod.params
plot_predictions(mod, by = ["Ta", "Tb"]).show()
avg_comparisons(mod, variables = "Ta",
  newdata = dat.filter(pl.col("Tb") == 0))
avg_comparisons(mod, variables=["Ta", "Tb"], cross=True)
avg_comparisons(mod, variables = "Ta", by = "Tb")
avg_comparisons(mod, variables = "Ta", by = "Tb",
  hypothesis = "b1 - b0 = 0")
```

10. Interactions and polynomials

Section 10.1

Categorical-by-categorical

```
dat = get_dataset("interaction_01").to_pandas()
mod = logit("Y ~ X * M", data=dat).fit()
avg_predictions(mod, by=["X", "M"])
plot_predictions(mod, by=["M", "X"]).show()
avg_comparisons(mod, variables = "X")
avg_comparisons(mod, variables = "X", by = "M")
avg_comparisons(mod,
  variables = "X",
  by = "M",
  hypothesis = "b2 - b0 = 0")
```

Categorical-by-continuous

```
dat = get_dataset("interaction_02")
mod = logit("Y ~ X * M", data=dat.to_pandas()).fit()
mod.summary()
fivenum = lambda x: np.quantile(x, [0, 0.25, 0.5, 0.75, 1])
range = lambda x: np.quantile(x, [0, 1])
p = predictions(mod, newdata=datagrid(X = [0, 1], M = fivenum))
plot_predictions(mod, condition = ["M", "X"]).show()
avg_comparisons(mod, variables = "X")
comparisons(mod, variables = "X", newdata=datagrid(M = range))
comparisons(mod, variables = "X",
  newdata=datagrid(M = range),
  hypothesis = "b1 - b0 = 0")
```

Continuous-by-continuous

```
dat = get_dataset("interaction_03")
mod = logit("Y ~ X * M", data=dat).fit()
predictions(mod, newdata = datagrid(
  X = [-2, 2],
  M = [-1, 0, 1]))
plot_predictions(mod, condition=["X", "M"]).show()
avg_slopes(mod, variables = "X")
slopes(mod, variables = "X", newdata = datagrid(M = fivenum))
plot_slopes(mod, variables = "X", condition = "M").show()
slopes(mod, variables = "X", newdata=datagrid(M = range))
slopes(mod, variables = "X", newdata=datagrid(M = range),
  hypothesis = "b1 - b0 = 0")
```

Multiple interactions

```
dat = get_dataset("interaction_04")
mod = logit("Y ~ X * M1 * M2", data = dat).fit()
mod.summary()
plot_predictions(mod, by = ["X", "M1", "M2"]).show()
avg_predictions(mod, newdata = datagrid(X = 0, M1 = 0, M2 = 0))
avg_comparisons(mod, variables = "X")
avg_comparisons(mod, variables = "X", by = "M1",
  hypothesis = "b1 - b0 = 0")
avg_comparisons(mod,
  variables = "X", by = ["M2", "M1"])
avg_comparisons(mod,
  hypothesis = "b1 - b0 = 0",
  variables = "X", by = ["M2", "M1"])
avg_comparisons(mod,
  hypothesis = "b3 - b2 = 0",
  variables = "X", by = ["M2", "M1"])
avg_comparisons(mod,
  hypothesis = "(b1 - b0) - (b3 - b2) = 0",
  variables = "X", by = ["M2", "M1"])
```

Section 10.2

```
dat = get_dataset("polynomial_01").to_pandas()
mod_linear = ols("Y ~ X", data = dat).fit()
plot_predictions(mod_linear, condition="X", points=0.05).show()
mod_cubic = ols("Y ~ X + I(X**2) + I(X**3)", data = dat).fit()
plot_predictions(mod_cubic, condition="X", points=0.05).show()
slopes(mod_cubic, variables="X", newdata=datagrid(X=[-2, 0, 2]))

dat = get_dataset("polynomial_02").to_pandas()
mod_cubic = ols("Y ~ X + I(X**2) + I(X**3)", data = dat).fit()
plot_predictions(mod_cubic, condition="X", points = 0.1).show()
mod_cubic_interaction = ols("Y ~ M * (X + I(X**2) + I(X**3))",
  data=dat).fit()
plot_predictions(mod_cubic_interaction,
  condition=["X", "M"], points=0.1).show()
fivenum = lambda x: np.quantile(x, [0, 0.25, 0.5, 0.75, 1])
slopes(mod_cubic_interaction, variables="X",
  newdata=datagrid(M=[0, 1], X=fivenum))
```

---

13. Machine learning

```
from marginaleffects import *
import numpy as np
import polars.selectors as cs
from sklearn.pipeline import make_pipeline
from sklearn.model_selection import train_test_split
from sklearn.preprocessing import (OneHotEncoder,
                                   FunctionTransformer)
from sklearn.compose import make_column_transformer
from xgboost import XGBRegressor

airbnb = get_dataset("airbnb")

airbnb[:5, :6]

train, test = train_test_split(airbnb, train_size = 3/4)
```

Section 13.2

```python
# Function to convert data frames into X and y matrices
def df_to_xy(data):
  y = data.select(cs.by_name("price", require_all=False))
  X = data.select(~cs.by_name("price", require_all=False))
  return y, X


# Convert categorical variables to dummies
catvar = airbnb.select(~cs.numeric()).columns
preprocessor = make_column_transformer(
  (OneHotEncoder(), catvar),
  remainder=FunctionTransformer(lambda x: x.to_numpy()),)


# Scikit-Learn pipeline: pre-process and fit
pipeline = make_pipeline(preprocessor, XGBRegressor())


# Run the pipeline
mod = fit_sklearn(
  df_to_xy,
  data=train,
  engine=pipeline,)

avg_predictions(mod, by = "unit_type", newdata = test)

plot_predictions(mod, by = ["bedrooms", "unit_type"]).show()

airbnb_subset = train.sample(10000)

grid = datagrid(
  bedrooms = np.unique,
  unit_type = np.unique,
  newdata = airbnb_subset,
  grid_type = "counterfactual")

plot_predictions(mod,
  by = ["bedrooms", "unit_type"],
  newdata = grid).show()
```

Section 13.3

```
avg_comparisons(mod, variables = {"bedrooms": 2})
avg_comparisons(mod,
  variables = ["bedrooms", "Wireless Internet"],
  cross = True)
```

Roadmap

At the time of writing (April 2025), the following features were not available in the Python version of **marginaleffects**. Please check marginaleffects.com for developments.

- Bayesian models.
- Mixed effects models.
- Categorical outcome models.
- Functions in the **comparison** argument of **comparisons()**.
- Rug plots in **plot_*()** functions.
- Bootstrap, simulation-based inference, and conformal prediction.
- Multiple comparison correction.
- Clustered standard errors.
- Robust standard errors and link scale predictions for logistic regression models with Statsmodels.

Bibliography

Alberto Abadie, Susan Athey, Guido W Imbens, and Jeffrey M Wooldridge. When should you adjust standard errors for clustering? *The Quarterly Journal of Economics*, 138(1):1–35, October 2022. ISSN 0033-5533. doi: 10.1093/qje/qjac038.

Rohan Alexander. *Telling Stories with Data: With Applications in R*. Chapman and Hall/CRC, 2023.

JJ Allaire, Yihui Xie, Jonathan McPherson, Javier Luraschi, Kevin Ushey, Aron Atkins, Hadley Wickham, Joe Cheng, Winston Chang, and Richard Iannone. *rmarkdown: Dynamic Documents for R*, 2022. URL https://cran.r-project.org/package=rmarkdown. R Package Version 2.19.

Anastasios N. Angelopoulos and Stephen Bates. A gentle introduction to conformal prediction and distribution-free uncertainty quantification, September 2022. URL http://arxiv.org/abs/2107.07511. arXiv:2107.07511 [cs, math, stat].

Joshua D Angrist and Jörn-Steffen Pischke. *Mostly harmless econometrics: An empiricist's companion*. Princeton University Press, 2009.

Vincent Arel-Bundock. modelsummary: Data and model summaries in R. *Journal of Statistical Software*, 103(1):1–23, 2022. doi: 10.18637/jss.v103.i01.

Vincent Arel-Bundock. *tinytable: Simple and Configurable Tables in 'HTML', 'LaTeX', 'Markdown', 'Word', 'PNG', 'PDF', and 'Typst' Formats*, 2025. URL https://vincentarelbundock.github.io/tinytable/. R package version 0.7.0.4.

Vincent Arel-Bundock, Noah Greifer, and Andrew Heiss. How to interpret statistical models using marginaleffects for R and Python. *Journal of Statistical Software*, 111(9):1–32, 2024. doi: 10.18637/jss.v111.i09. URL https://marginaleffects.com.

Peter M. Aronow and Benjamin T. Miller. *Foundations of Agnostic Statistics*. Cambridge University Press, Cambridge, 2019. ISBN 9781316831762. doi: 10.1017/9781316831762. URL https://doi.org/10.1017/9781316831762.

Kirk Bansak. Estimating causal moderation effects with randomized treatments and non-randomized moderators. *Journal of the Royal Statistical Society: Series A*, 184(1):65–86, 2021. doi: 10.1111/rssa.12614.

Douglas Bates and Dirk Eddelbuettel. Fast and elegant numerical linear algebra using the RcppEigen package. *Journal of Statistical Software*, 52(5): 1–24, 2013. doi: 10.18637/jss.v052.i05.

Douglas Bates, Martin Mächler, Ben Bolker, and Steve Walker. Fitting linear mixed-effects models using lme4. *Journal of Statistical Software*, 67(1):1–48, 2015. doi: 10.18637/jss.v067.i01.

Laurent Bergé. Efficient estimation of maximum likelihood models with multiple fixed-effects: the R package FENmlm. *CREA Discussion Papers*, (13), 2018.

Roger Berger and George Casella. *Statistical Inference*. CRC Press, 2 edition, 2024. ISBN 9781003456285.

Bernd Bischl, Raphael Sonabend, Lars Kotthoff, and Michel Lang, editors. *Applied Machine Learning Using mlr3 in R*. Chapman and Hall/CRC, USA, 1st edition, January 2024. ISBN 978-1032507545.

Graeme Blair, Jasper Cooper, Alexander Coppock, Macartan Humphreys, and Luke Sonnet. *estimatr: Fast Estimators for Design-Based Inference*, 2024. URL https://declaredesign.org/r/estimatr/. R package version 1.0.4, https://github.com/DeclareDesign/estimatr.

Thomas Brambor, William Roberts Clark, and Matt Golder. Understanding interaction models: Improving empirical analyses. *Political analysis*, 14(1): 63–82, 2006.

Mollie E. Brooks, Kasper Kristensen, Koen J. van Benthem, Arni Magnusson, Casper W. Berg, Anders Nielsen, Hans J. Skaug, Martin Maechler, and Benjamin M. Bolker. glmmTMB balances speed and flexibility among packages for zero-inflated generalized linear mixed modeling. *The R Journal*, 9(2):378–400, 2017. doi: 10.32614/RJ-2017-066.

Gábor Békés and Gábor Kézdi. *Data Analysis for Business, Economics, and Policy*. Cambridge University Press, Cambridge, UK, 2021. ISBN 978-1108716208. URL https://www.gabors-data-analysis.com.

Paul C. Bürkner. *The brms Book: Applied Bayesian Regression modeling Using R and Stan*. Chapman and Hall/CRC, 2024. URL https://paulbuerkner.com/software/brms-book.

Paul-Christian Bürkner. brms: An R package for Bayesian multilevel models using Stan. *Journal of Statistical Software*, 80(1):1–28, 2017. doi: 10.18637/jss.v080.i01.

A Colin Cameron and Pravin K Trivedi. *Microeconometrics: methods and applications*. Cambridge University Press, 2005.

Angelo Canty and B. D. Ripley. *boot: Bootstrap R (S-PLUS) Functions*, 2022. URL https://cran.r-project.org/package=boot. R Package Version 1.3-28.1.

Arthur Chatton and Julia M. Rohrer. The causal cookbook: Recipes for propensity scores, g-computation, and doubly robust standardization. *Advances in Methods and Practices in Psychological Science*, 7(1):25152459241236149, January 2024. ISSN 2515-2459. doi: 10.1177/25152459241236149.

Carlos Cinelli, Andrew Forney, and Judea Pearl. A crash course in good and bad controls. *Sociological Methods & Research*, 53(3):1071–1104, 2024.

William Roberts Clark and Matt Golder. *Interaction Models: Specification and Interpretation*. Methodological Tools in the Social Sciences. Cambridge University Press, 2023.

Gábor Csárdi and Hannes Mühleisen. *nanoparquet: Read and Write 'Parquet' Files*, 2024. URL https://CRAN.R-project.org/package=nanoparquet. R package version 0.3.1.

Philip A. Dawid. Probability, symmetry and frequency. *The British Journal for the Philosophy of Science*, 36(2):107–128, 1985.

Peng Ding. *A First Course in Causal Inference*. Chapman & Hall, 1st edition, 2024. ISBN 9781032758626.

Tiffany Ding, Anastasios N. Angelopoulos, Stephen Bates, Michael I. Jordan, and Ryan J. Tibshirani. Class-conditional conformal prediction with many classes, June 2023. URL http://arxiv.org/abs/2306.09335. arXiv:2306.09335 [cs, stat].

Matt Dowle and Arun Srinivasan. *data.table: Extension of data.frame*, 2022. URL https://r-datatable.com.

Dirk Eddelbuettel, Romain Francois, JJ Allaire, Kevin Ushey, Qiang Kou, Nathan Russell, Iñaki Ucar, Doug Bates, and John Chambers. *Rcpp: Seamless R and C++ Integration*, 2025. URL https://CRAN.R-project.org/package=Rcpp. R package version 1.0.14.

Bradley Efron and R.J. Tibshirani. *An Introduction to the Bootstrap*. Chapman and Hall/CRC, New York, 1994. ISBN 9780429246593. doi: 10.1201/9780429246593.

Ray C. Fair. A theory of extramarital affairs. *Journal of Political Economy*, 86:45–61, 1978.

W Holmes Finch, Jocelyn E Bolin, and Ken Kelley. *Multilevel modeling using R*. Chapman and Hall/CRC, 2019.

David A. Freedman. On regression adjustments to experimental data. *Advances in Applied Mathematics*, 40(2):180–193, February 2008.

Hannah Frick, Fanny Chow, Max Kuhn, Michael Mahoney, Julia Silge, and Hadley Wickham. *rsample: General Resampling Infrastructure*, 2022. URL https://rsample.tidymodels.org.

Andrew Gelman and Jennifer Hill. *Data Analysis Using Regression and Multilevel/Hierarchical Models.* Cambridge University Press, 1st edition, 2006. ISBN 978-0521686891. URL http://www.stat.columbia.edu/~gelman/arm/.

Andrew Gelman, John B. Carlin, Hal S. Stern, David B. Dunson, Aki Vehtari, and Donald B. Rubin. *Bayesian Data Analysis.* Chapman and Hall/CRC, New York, 3rd edition, 2013. ISBN 9780429113079. doi: 10.1201/b16018. URL https://doi.org/10.1201/b16018. eBook published 5 July 2015.

Andrew Gelman, Aki Vehtari, Daniel Simpson, Charles C. Margossian, Bob Carpenter, Yuling Yao, Lauren Kennedy, Jonah Gabry, Paul-Christian Bürkner, and Martin Modrák. Bayesian workflow, 2020. URL https://arxiv.org/abs/2011.01808.

Alan Genz and Frank Bretz. *Computation of Multivariate Normal and t Probabilities.* Lecture Notes in Statistics. Springer-Verlag, Heidelberg, 2009. ISBN 978-3-642-01688-2.

Paul Goldsmith-Pinkham, Peter Hull, and Michal Kolesár. Contamination bias in linear regressions. *American Economic Review*, Forthcoming. ISSN 0002-8282. Forthcoming.

Noah Greifer. *fwb: Fractional Weighted Bootstrap*, 2022. URL https://ngreifer.github.io/fwb/.

Noah Greifer and Elizabeth A Stuart. Choosing the estimand when matching or weighting in observational studies. *arXiv e-prints*, pages arXiv–2106, 2021.

Noah Greifer, Steven Worthington, Stefano Iacus, and Gary King. *clarify: Simulation-Based Inference for Regression Models*, 2024. URL https://CRAN.R-project.org/package=clarify. R package version 0.2.1.

Michael P Grosz, Julia M Rohrer, and Felix Thoemmes. The taboo against explicit causal inference in nonexperimental psychology. *Perspectives on Psychological Science*, 15(5):1243–1255, 2020.

Jens Hainmueller, Jonathan Mummolo, and Yiqing Xu. How much should we trust estimates from multiplicative interaction models? simple tools to improve empirical practice. *Political Analysis*, 27(2):163–192, 2019.

Bruce Hansen. *Probability and Statistics for Economists.* Princeton University Press, Princeton, NJ, 1 edition, August 2022a. ISBN 9780691235899. URL https://press.princeton.edu/books/hardcover/9780691235899/probability-and-statistics-for-economists.

Bruce Hansen. *Econometrics.* Princeton University Press, Princeton, NJ, 1st edition, August 2022b. ISBN 9780691223248. URL https://press.princeton.edu/books/hardcover/9780691223248/econometrics. Published in the US on August 16, 2022, and in the UK on October 11, 2022.

S. N. Hansen and M. Overgaard. Variance estimation for average treatment effects estimated by g-computation. *Metrika*, 2024. doi: 10.1007/s00184-024-00962-4.

Frank Harrell. Statistical thinking - avoiding one-number summaries of treatment effects for rcts with binary outcomes, June 2021. URL https://www.fharrell.com/post/rdist/. Accessed: 2024-07-24.

Andrew Heiss. Marginalia: A guide to figuring out what the heck marginal effects, marginal slopes, average marginal effects, marginal effects at the mean, and all these other marginal things are, 2022. URL https://www.andrewheiss.com/blog/2022/05/20/marginalia.

Lionel Henry and Hadley Wickham. *rlang: Functions for Base Types and Core R and tidyverse Features*, 2023. URL https://rlang.r-lib.org.

Miguel A Hernán. The c-word: scientific euphemisms do not improve causal inference from observational data. *American journal of public health*, 108 (5):616–619, 2018.

Miguel A Hernán and James M Robins. *Causal Inference: What If*. Chapman & Hall/CRC, Boca Raton, December 2020.

Jennifer Hill, George Perrett, Stacey Hancock, Le Win, and Yoav Bergner. Causal language and statistics instruction: Evidence from a randomized experiment. *STATISTICS EDUCATION RESEARCH JOURNAL*, 23(1), August 2024. ISSN 1570-1824. doi: 10.52041/serj.v23i1.673.

James S. Hodges. *Richly Parameterized Linear Models: Additive, Time Series, and Spatial Models Using Random Effects*. Chapman & Hall, 1st edition, 2021. ISBN 9780367533731. Originally published in 2014.

Allison Marie Horst, Alison Presmanes Hill, and Kristen B Gorman. *palmerpenguins: Palmer Archipelago (Antarctica) penguin data*, 2020. URL https://allisonhorst.github.io/palmerpenguins/. R package version 0.1.0.

Rob J Hyndman and George Athanasopoulos. *Forecasting: principles and practice*. OTexts, 2018.

Kosuke Imai, Luke Keele, Dustin Tingley, and Teppei Yamamoto. Unpacking the black box of causality: Learning about causal mechanisms from experimental and observational studies. *American Political Science Review*, 105 (4):765–789, 2011.

Guido W Imbens and Donald B Rubin. *Causal inference in statistics, social, and biomedical sciences*. Cambridge university press, 2015.

Guido W Imbens, Donald B Rubin, and Bruce I Sacerdote. Estimating the effect of unearned income on labor earnings, savings, and consumption:

Evidence from a survey of lottery players. *American economic review*, 91(4): 778–794, 2001.

Gareth James, Daniela Witten, Trevor Hastie, and Robert Tibshirani. *An Introduction to Statistical Learning: with Applications in R*. Springer, second edition, 2021. ISBN 978-1071614174. URL https://www.springer.com/gp/book/9781071614174.

Gareth James, Daniela Witten, Trevor Hastie, and Robert Tibshirani. *An Introduction to Statistical Learning: with Applications in Python*. Springer Nature, USA, 2023rd edition, July 2023. ISBN 978-3031387463.

Daniel Kahneman and Amos Tversky. Prospect theory: An analysis of decision under risk. *Econometrica*, 47(2):263–291, 1979. ISSN 00129682, 14680262. URL http://www.jstor.org/stable/1914185.

Cindy D. Kam and Robert J. Franzese, Jr. *Modeling and Interpreting Interactive Hypotheses in Regression Analysis*. University of Michigan Press, 2009. ISBN 9780472069699. doi: 10.3998/mpub.206871. URL https://doi.org/10.3998/mpub.206871.

Matthew Kay. *ggdist: Visualizations of Distributions and Uncertainty*, 2023. URL https://mjskay.github.io/ggdist/. R Package Version 3.2.1.

Luke Keele, Randolph T. Stevenson, and Felix Elwert. The causal interpretation of estimated associations in regression models. *Political Science Research and Methods*, 8(1):1–13, January 2020. ISSN 2049-8470, 2049-8489. doi: 10.1017/psrm.2019.31.

Gary King, Michael Tomz, and Jason Wittenberg. Making the most of statistical analyses: Improving interpretation and presentation. *American Journal of Political Science*, pages 347–361, 2000.

I. Krinsky and A. L. Robb. On approximating the statistical properties of elasticities. *Review of Economics and Statistics*, 68(4):715–719, 1986.

Max Kuhn and Julia Silge. *Tidy Modeling with R: A Framework for Modeling in the Tidyverse*. O'Reilly Media, USA, 1st edition, 2022. ISBN 978-1492096481.

Max Kuhn and Hadley Wickham. *Tidymodels: a collection of packages for modeling and machine learning usin g tidyverse principles.*, 2020. URL https://www.tidymodels.org.

Daniël Lakens, Anne M Scheel, and Peder M Isager. Equivalence testing for psychological research: A tutorial. *Advances in methods and practices in psychological science*, 1(2):259–269, 2018.

Michel Lang. checkmate: Fast argument checks for defensive R programming. *The R Journal*, 9(1):437–445, 2017. doi: 10.32614/RJ-2017-028.

Michel Lang, Martin Binder, Jakob Richter, Patrick Schratz, Fl orian Pfisterer, Stefan Coors, Quay Au, Giuseppe Casalicchio, Lars Kotthoff, and Bernd Bischl. mlr3: A modern object-oriented machine learning framework in R. *Journal of Open Source Software*, dec 2019. doi: 10.21105/joss.01903. URL https://joss.theoj.org/papers/10.21105/joss.01903.

Russell V. Lenth. *emmeans: Estimated Marginal Means, aka Least-Squares Means*, 2024. URL https://cran.r-project.org/package=emmeans. R Package Version 1.10.3.

Winston Lin. Agnostic notes on regression adjustments to experimental data: Reexamining freedman's critique. *Annals of Applied Statistics*, 7(1):295–318, March 2013. doi: 10.1214/12-AOAS583.

Daniel Lüdecke, Philip Waggoner, and Dominique Makowski, "insight: A Unified Interface to Access Information from Model Objects in R," *Journal of Open Source Software* 4, no. 38 (2019): 1412, https://doi.org/10.21105/joss.01412.

Ian Lundberg, Rebecca Johnson, and Brandon M. Stewart. What is your estimand? defining the target quantity connects statistical evidence to theory. *American Sociological Review*, 86(3):532–565, June 2021. ISSN 0003-1224. doi: 10.1177/00031224211004187.

Dominic Magirr, Craig Wang, Alexander Przybylski, and Mark Baillie. Estimating the variance of covariate-adjusted estimators of average treatment effects in clinical trials with binary endpoints. December 2024. doi: 10.31219/osf.io/k56v8. URL https://osf.io/k56v8_v1.

Richard McElreath. *Statistical Rethinking: A Bayesian Course with Examples in R and STAN*. Chapman and Hall/CRC, New York, 2nd edition, 2020. ISBN 9780429029608. doi: 10.1201/9780429029608. URL https://doi.org/10.1201/9780429029608. eBook published 15 March 2020.

Wes McKinney. *Python for data analysis*. " O'Reilly Media, Inc.", 2022.

Alex Molak. Causal bandits podcast, March 2024. URL https://causalbandit spodcast.com/. Interview with Stephen Senn. Episode Duration: 50m00s.

Stephen L Morgan and Christopher Winship. *Counterfactuals and causal inference*. Cambridge University Press, 2015.

Kirill Müller. *here: A Simpler Way to Find Your Files*, 2020. URL https://here.r-lib.org/. R package version 1.0.1, https://github.com/r-lib/here.

Raymond S. Nickerson. *Cognition and Chance: The Psychology of Probabilistic Reasoning*. Psychology Press, 1st edition, 2004. doi: 10.4324/9781410610836. URL https://doi.org/10.4324/9781410610836.

Joseph T. Ornstein. Getting the most out of surveys: Multilevel regression and poststratification. In Alessia Damonte and Fedra Negri, editors, *Causality in Policy Studies: A Pluralist Toolbox*, pages 99–122. Springer, 2023. doi: 10.1007/978-3-031-12982-7_5.

Judea Pearl. *Causality*. Cambridge university press, 2009.

Judea Pearl. Interpretation and identification of causal mediation. *Psychological Methods*, 19(4):459–481, December 2014. doi: 10.1037/a0036434.

Judea Pearl and Dana Mackenzie. *The book of why: the new science of cause and effect*. Basic books, 2018.

Thomas Lin Pedersen. *patchwork: The Composer of Plots*, 2024. URL https://patchwork.data-imaginist.com. R package version 1.3.0, https://github.com/thomasp85/patchwork.

F. Pedregosa, G. Varoquaux, A. Gramfort, V. Michel, B. Thirion, O. Grisel, M. Blondel, P. Prettenhofer, R. Weiss, V. Dubourg, J. Vanderplas, A. Passos, D. Cournapeau, M. Brucher, M. Perrot, and E. Duchesnay. Scikit-learn: Machine learning in Python. *Journal of Machine Learning Research*, 12: 2825–2830, 2011.

James Pustejovsky. *clubSandwich: Cluster-Robust (Sandwich) Variance Estimators with Small-Sample Corrections*, 2023. URL https://CRAN.R-project.org/package=clubSandwich. R package version 0.5.10.

Carlisle Rainey. Arguing for a negligible effect. *American Journal of Political Science*, 58(4):1083–1091, 2014.

Carlisle Rainey. A careful consideration of clarify: simulation-induced bias in point estimates of quantities of interest. *Political Science Research and Methods*, 12(3):614–623, 2024. doi: 10.1017/psrm.2023.8.

R Core Team. *R: A Language and Environment for Statistical Computing*. R Foundation for Statistical Computing, Vienna, Austria, 2022. URL https://www.R-project.org/.

Neal Richardson, Ian Cook, Nic Crane, Dewey Dunnington, Romain François, Jonathan Keane, Dragoș Moldovan-Grünfeld, Jeroen Ooms, Jacob Wujciak-Jens, and Apache Arrow. *arrow: Integration to 'Apache' 'Arrow'*, 2025. URL https://CRAN.R-project.org/package=arrow. R package version 18.1.0.1.

Skipper Seabold and Josef Perktold. statsmodels: Econometric and statistical modeling with Python. In *Proceedings of the 9th Python in Science Conference*, volume 57,61, pages 10–25080. Austin, TX, 2010.

Rebecca L Thornton. The demand for, and impact of, learning hiv status. *American Economic Review*, 98(5):1829–1863, 2008.

John W. Tukey. *Exploratory Data Analysis*. Addison-Wesley, Reading, MA, 1977.

Tyler J. VanderWeele. On the distinction between interaction and effect modification. *Epidemiology*, 20(6):863–871, 2009. doi: 10.1097/EDE.0b013e 3181ba333c.

W. N. Venables and B. D. Ripley. *Modern Applied Statistics with S*. Springer, New York, fourth edition, 2002a. URL https://www.stats.ox.ac.uk/pub/M ASS4/. ISBN 0-387-95457-0.

W. N. Venables and B. D. Ripley. *Modern Applied Statistics with S*. Springer, New York, fourth edition, 2002b. URL https://www.stats.ox.ac.uk/pub/M ASS4/. ISBN 0-387-95457-0.

Larry Wasserman. *All of Statistics: A Concise Course in Statistical Inference*. Springer Texts in Statistics. Springer, New York, NY, 2004. ISBN 978-0-387-40272-7. doi: 10.1007/978-0-387-21736-9. URL https://link.springer.co m/book/10.1007/978-0-387-21736-9.

Larry Wasserman. *All of nonparametric statistics*. Springer Texts in Statistics. Springer, New York, NY, 2006.

Stefan Wellek. *Testing Statistical Hypotheses of Equivalence and Noninferiority*. CRC Press, 2010. ISBN 9781439808184. URL https://www.taylorfrancis.co m/books/mono/10.1201/EBK1439808184/testing-statistical-hypotheses-equivalence-noninferiority-stefan-wellek.

Daniel Westreich and Sander Greenland. The table 2 fallacy: Presenting and interpreting confounder and modifier coefficients. *American Journal of Epidemiology*, 177(4):292–298, 2013. doi: 10.1093/aje/kws412.

Hadley Wickham. *ggplot2: Elegant Graphics for Data Analysis*. Springer-Verlag New York, 2016. ISBN 978-3-319-24277-4. URL https://ggplot2.tidy verse.org.

Hadley Wickham, Mine Çetinkaya Rundel, and Garrett Grolemund. *R for Data Science: Import, Tidy, Transform, Visualize, and Model Data*. O'Reilly Media, Sebastopol, CA, 2 edition, 2023. ISBN 978-1492097402. URL https://www.amazon.ca/dp/1492097403. Available in Kindle and Paperback editions.

Yihui Xie. *knitr: A General-Purpose Package for Dynamic Report Generation in R*, 2025. URL https://yihui.org/knitr/. R package version 1.50.

T. Ye, M. Bannick, Y. Yi, and J. Shao. Robust variance estimation for covariate-adjusted unconditional treatment effect in randomized clinical trials with binary outcomes. *Statistical Theory and Related Fields*, 7(2):159–163, 2023.

Achim Zeileis and Yves Croissant. Extended model formulas in R: Multiple parts and multiple responses. *Journal of Statistical Software*, 34(1):1–13, 2010. doi: 10.18637/jss.v034.i01.

Achim Zeileis, Susanne Köll, and Nathaniel Graham. Various versatile variances: An object-oriented implementation of clustered covariances in R. *Journal of Statistical Software*, 95(1):1–36, 2020. doi: 10.18637/jss.v095.i01. URL https://cran.r-project.org/package=sandwich.

Jinhui Zhao, Tim Stockwell, Tim Naimi, Sam Churchill, James Clay, and Adam Sherk. Association Between Daily Alcohol Intake and Risk of All-Cause Mortality: A Systematic Review and Meta-analyses. *JAMA Network Open*, 6(3):e236185–e236185, 03 2023. ISSN 2574-3805. doi: 10.1001/jamanetworko pen.2023.6185. URL https://doi.org/10.1001/jamanetworkopen.2023.6185.

Index

For Product Safety Concerns and Information please contact our EU
representative GPSR@taylorandfrancis.com
Taylor & Francis Verlag GmbH, Kaufingerstraße 24, 80331 München, Germany